Designing Service Excellence

People and Technology

Brian Hunt

Toni Ivergård

CRC Press
Taylor & Francis Group
Boca Raton London New York

CRC Press is an imprint of the
Taylor & Francis Group, an **informa** business

CRC Press
Taylor & Francis Group
6000 Broken Sound Parkway NW, Suite 300
Boca Raton, FL 33487-2742

First issued in hardback 2019

© 2015 by Taylor & Francis Group, LLC
CRC Press is an imprint of Taylor & Francis Group, an Informa business

No claim to original U.S. Government works

ISBN-13: 978-1-4398-4046-7 (hbk)

Library of Congress Cataloging-in-Publication Data

Hunt, Brian.
  Designing service excellence : people and technology / Brian Hunt and Toni Ivergård.
    pages cm
  Includes bibliographical references and index.
  ISBN 978-1-4398-4046-7 (alk. paper)
    1. Customer services. 2. Customer services--Technological innovations. I. Ivergård, Toni II. Title.

HF5415.5.H867 2015
658.8'12--dc23
                                                                                2014023195

Visit the Taylor & Francis Web site at
http://www.taylorandfrancis.com

and the CRC Press Web site at
http://www.crcpress.com

Dedication

The authors gratefully acknowledge their debt to the late Richard Normann, whose writings and scholarship have stimulated our thinking over many years. We also acknowledge the work of Jan Carlzon, especially his insightful ideas in his classic book *Moments of Truth* (Cambridge, MA: Ballinger Publishing).

Contents

Preface

This is a book about service excellence. The perspectives we take are people, technology, and people *and* technology. We identify service (in a variety of forms) as an area of business and management where rapid change is taking place. We believe our chosen perspectives are among the key drivers of change. Social, cultural, and technological developments are influencing the ways in which customers contact, negotiate, and purchase services from their chosen service providers. These same developments are also driving communications between customers relating to the services they buy and are willing to recommend to others (or otherwise). Intermingled, these features of our current-day lives have changed the nature of service provision and service use. Not only are such changes continuing, but in certain dimensions they appear to be accelerating. Always somewhat precarious, balances of power have been shifting for some time. Customers (service users) have become more techno-savvy and thereby may have wider access to information about the services they use. Communities of users employ technology to describe and evaluate service provided as well as to draw up wish lists of services that should be provided. As we discuss throughout our book, communities of users can be co-creators of service offerings. However, in order for such communities to be developed, service providers need to be aware of the mutual benefits as well as establishing processes and mechanisms whereby users become part of their community. For service providing organizations, there are three fundamental issues: awareness of the service from users' perspectives, sensitivity to the contribution that users can make to service development and improvements, and the creative design of engagement processes with service users. Arguably, the latter task is the most important and also the most difficult to achieve. A hallmark of success would be that service users feel that their service provider has that all-elusive "wow" factor. A hallmark of failure would be when end users do not really care to become part of a service provider's community as they are less than satisfied with the service, see other users defecting to rival service providers, and are actively seeking to use services provided by a competitor. Myriad examples of this scenario can be seen in the airline, banking, credit card, fashion, mobile phone, and travel industries. In these highly competitive markets, the service provided by some organizations can politely be described as patchy. Service inconsistency (where the user cannot wholly predict the quality of future service from the ambiance of the current service) is not a sustainable business model for an enterprise. A wide gap between service excellence on one occasion and service mediocrity on a subsequent occasion tends to deflate a user's opinion, which in turn reduces expectations of the service. When service users have reduced expectations, especially of a service provider's ability to replicate the quality provided on a previous occasion, one or more competitors may seize the opportunity to "poach" the customer with a more attractive service offering. In industries where customer loyalty is fragile, which tends to be most industries, it is a truism to state, "You are only as good as your last service." The rationale here is that a customer tends to remember the last occasion on which he or she received service (be it good, bad, or ugly), and that experience colors future willingness to use the same service provider. Restaurateurs in particular realize that poor service on the most recent occasion on which the customer came to dine is likely to reduce future business from that customer and from his or her friends and acquaintances. There tend to be legitimate business reasons why some enterprises offer a free meal, free room, or free upgrade when a service encounter goes bad. Without the offer of a free next time, the customer would likely never return.

Over the past decades, developments in technology, especially information and communications technology (ICT), have driven changes in business environments. The so-called information age or digital age accelerates the transmission of information across continents with one click of a mouse. But the same infrastructure of transmission can also disseminate opinion, rumor, hearsay, conjecture, and sometimes outright misinformation and falsehoods. Determining what information is real or fabricated brings problems for service users, especially first-time service users, and service providers, especially in the complex environment of communication with potential customers. In our research and as reported in our book, we have focused on how people interact with technology in service provision. In this we consider what have become traditional modes of service interaction as well as modes of interactions that continue to evolve.

It has become a cliché to state that technology is changing our lives. We feel it is a misconception to see this as a feature of our modern age. In fact, technology throughout history changed people's lives. Advancements in heating and lighting (the discovery of fire) and farming began life-changing patterns that have endured for millennia. Inventions that were seen as fads or clunking novelties (the wheel, the cart, the steam engine, the horseless carriage, the typewriter, and the mobile phone, to name but a few) themselves developed into machines that modern societies cannot do without. It is somewhat intriguing to consider that the technologies that support service may be at a very early stage of development. If that is indeed the case, then some aspects of service technology have a long way to go before anything near perfection can be achieved.

Our third perspective on service by people and technology is the moment of truth. A metaphor taken from bullfighting, the moment of truth is the point at which service is offered and delivered. This is where the customer perceives service quality, for better or worse. From this point onward, the customer and the service provider co-engage to produce the service outcomes. In the nature of service (especially service delivered and used face-to-face) this is unavoidable. Astute service organizations engage with the moment of truth to seek ways of developing their service. Organizations that have known past success may become complacent about service quality and pay less attention to the opinions of customers. Conversely, service organizations that are striving for success may need to try harder (to quote the marketing tagline of Avis). This is especially so in a competitive market where one or two "name" organizations set standards for the industry that smaller, less experienced, or newer organizations attempt to emulate.

Service can be considered from the perspective of supply ("we produce service, we expect customers to want what we produce") or demand ("we produce service according to our perceptions of customer needs and aspire to satisfy these needs"). When an organization bases its business model on service supply rather than service demand, executives and managers may not set a high priority on customer feedback. This is short-term thinking. At the best of times, it is not easy to design, manage, and deliver service and to deliver consistently high levels of service excellence. When technologies mediate in the service process, executives and managers face more challenges in service design and execution. Service is part of everyday social interactions. Service with no apparent human service provider changes a customer's perceptions of the service as well as the framework for service management. In essence, the tasks of managing the service become more difficult. Engagement with service users is less straightforward, and both customer and the service provider see the service encounter differently, that is, from differing perspectives as well as differing perceptions of their roles. Technology outperforms humans, but at least at the present time, humans perform better on a range of emotions. We all are aware from personal experiences that technology-aided services can be time-consuming as well as a strain on our patience. Service "with a human face" can deal with misapprehensions and instances of miscommunications. A human service provider can sense the direction the service is taking (e.g., toward usage or nonusage) and react accordingly. In the presence of a human service provider, a service user can engage emotionally within the service encounter and feel the social nature of the service provision.

We have personally experienced and received reports from others of service in several continents and in cities as far flung and diverse as Amsterdam, Bahrain, Bangkok, Berne, Chicago, Copenhagen, Dubai, Frankfurt, Jeddah, Kuala Lumpur, Las Vegas, Lucerne, Milan, Osaka, Oslo, Paris, Rome, San Francisco, Seattle, Singapore, Stockholm, The Hague, Tokyo, Vancouver, Vienna, and Zurich. In each of these locales we have noted various aspects of service. We are increasingly aware that service excellence is not necessarily confined to carpeted, high-priced, global brands where training budgets are high and employees are recruited with high levels of experience and then receive further training. In such environments, executives are willing to invest generously per capita in their employees. In such environments investments can also be made in recruitment and psychometric testing to ensure that (at least on paper) the organization is hiring the best and the brightest. However, even with rigorous screening in place, it is not uncommon for service delivery to go awry. And, for some reasons, high-value brands have more than their fair quota of front-line employees who can deliver service with a sneer. When brand value is high, there is much to lose from perceived wide gaps between price premium and value delivered via service. Disappointing levels of service can be a feature of multinational corporations that use glossy print media to promote the values and virtues of their service.

Our personal experience and verbal and written reports from colleagues indicate that excellent service can be found in one-man or one-woman enterprises in the meaner streets of town. Undoubtedly, when survival is on the line, service becomes crucial. In these business environments service providers seem willing and able to "go the extra mile" to gain and retain a customer. This especially applies to business owners (mom-and-pop stores, for example), or family businesses where revenues support an extended family. Clearly, there is not a one-to-one correlation between amount of budget spend and service excellence. We are reminded of an adage from management guru Peter Drucker that what is important is the value that a customer gets out of a service, not what the service provider puts in to it. This seems to us to convey the essence of service.

In various venues and locations where service encounters take place we have been pleasantly surprised. Whether coincidental or not, excellent service has tended to come when least expected: perhaps from gruff-looking taxi drivers at the end of their shift, from busy waiting staff in a crowded restaurant, from cabin crew on a packed long-haul flight, from desk officers in busy police stations, and from harried nursing staff in an A&E ward. Often, we and others have been amazed at the efforts of service providers who go the extra mile: the Qantas flight attendant on a Sydney–Bangkok flight who exchanged her own allotted meal for a passenger when his own selection was not available, the taxi driver in Jeddah who drove around without extra charge to find the right hotel, the receptionist in Singapore who instantly gave a room upgrade when the first room hadn't yet been cleaned. We also recall nonservice people who provided service with no expectation of reward, such as the passerby in late night Osaka who walked his pedal bicycle seven blocks to show the way to the hotel. These instances contrast sharply with service that could have been more polite and less mercenary: the bus company employee at Hong Kong who abruptly said, "Get change!" to a newly arrived passenger, the receptionist at the five-star hotel in Singapore whose manner stood in poor comparison to the bellboy, the hotel GM in Kuala Lumpur who seemed too distracted by arriving guests to deal with the complaint of a current guest.

Thus, we have had more than our share of disappointments. Descriptions of service from reports by our informants suggest that this emotion is not unusual. As we write in one of our chapters, customers seem to be irritated more by inconsistent service than by consistently poor service. In service, expectations matter. When receiving substandard service, we have engaged service providers in what seasoned members of the diplomatic community call a "full and frank exchange of views." We understand that some of our respondents also take this approach. We also know that the overwhelming majority of our respondents, who seem by no means shy and reserved, often do not complain about poor service. Research indicates that less than 5 percent of disappointed customers complain about service when it is inadequate. The vast majority of disappointed customers leave and don't return. In this context we conjecture that walking away is a form of complaint, albeit one that denies the service provided of valuable feedback. Perhaps that too is another form of registering a complaint.

Over a number of years, and in writing our book, we have enjoyed conversations with a number of colleagues. We have learned much from them and are delighted to be able to thank them here. On subjects connected with public management, we have exchanged our ideas with Anders L. Johansson, sometime director general of the Swedish Board of Labour, and Dr. Somphoch Nophakoon, former deputy secretary general of the Thai Office of the Civil Service Commission (OCSC) in Thailand. In the field of ergonomics we have had interesting talks with John Wood, co-founder of CCD Design & Ergonomics. Our horizons on service and marketing have been broadened by discussions with Dr. Nigel Bairstow, Dr. Nigel Barrett, Professor George Moschis, Professor Paul Patterson, Dr. Randall Shannon, and Dr. Gerard Tocquer. On management issues we have learned from our conversations with David Cramond, Dr. Astrid Kainzbauer, Dr. Roy Kowenberg, Michel Le Quellec, and Dr. Detlef Reis. We thank these colleagues for their willingness to share their ideas and help us develop our own. We accept that where mistakes have been made, they are our own. Jatupol Chawapatnakul greatly helped us with figures and diagrams.

At CRC Press/Taylor & Francis we owe much thanks to the support and unswerving patience of Cindy Carelli, Judith Simon, and Jessica Vakili. Without their encouragement this book would have taken us even longer to complete. While our procrastinations may have stretched their patience, we hope they feel this book is worth the wait.

We have been researching and writing this book for several years. During this time we have tried the patience of our respective wives, Mallika and Duangjai. We thank them both for their continued patience.

Brian Hunt
Bangkok, Thailand

Toni Ivergård
Stockholm, Sweden

About the Authors

Brian Hunt is a graduate of the Universities of East Anglia (UEA), Reading, and Bath, all in the UK. He has conducted business research projects at the Management School of the University of Bath, City University Business School, and at Imperial College, London. His most recent academic position was at the College of Management Mahidol University (CMMU) in Thailand where he was Director of Research (from 2000–2004) and Assistant Dean for Quality Assurance (from 2011–2014).

Earlier in his career, Brian worked in the training departments of Petromin (the Saudi Ministry of Petroleum) in Jeddah and the Petromin-Shell Refinery Corporation in Jubail, Saudi Arabia and managed national development projects funded by the UK government in Thailand and Malaysia.

Dr. Hunt has published widely in the areas of business and management, specifically in corporate strategy, learning in organizations, public management, and ergonomic approaches to management and organizational design. His most recent book (co-authored with Toni Ivergård) is the *Handbook of Control Room Design and Ergonomics: A Perspective for the Future* (published by CRC Press in 2008).

Brian attained his PhD in management from the University of Technology Sydney (UTS), Australia. In June 2014 he completed the prestigious Advanced Management and Leadership programme at the Saïd Management School of the University of Oxford.

Toni K.B. Ivergård has a postdoctorate in science of work from the Royal Institute of Technology, Sweden, a PhD in human sciences, and a master in ergonomics and cybernetics from Loughborough University, UK. He was a regional director of the Swedish National Research Institute of Working Life, and he has published numerous books and over 220 papers and articles. Currently he is director of a master of management in innovation and entrepreneurship at Rangsit University in Bangkok. He is also the managing director of the Ivergård Management Consultancy (IMC) Ltd. in Stockholm. IMC specializes in top leadership development.

Over several decades he has been in management positions in both the public and private sectors. He was R&D director of the Scandinavian Airline

University, managing director of the consultancy and research company ERGOLAB, and head of the environmental laboratory of the Cooperative Union and Wholesale Society. Recently he has focused on "learning at work," macro-aspects of HR and technology, and the role of corporations in society.

1

Service: Definitions and Attributes

Toward a Definition of Service

As a core of this book considers the provision and management of service, we believe it helpful at this early stage to provide one or more definitions of service and to discuss these. Service use is often immediate and (with few exceptions) is directly experienced by the user. Conventional examples of services include personal services such as a manicure, a haircut, and dental care, all of which are provided to the customer in that customer's presence. By way of contrast, personal items such as nail extensions, hairpieces, wigs and toupees, and dentures can have specialist work done to them in the absence of the user. Hair extensions, wigs, and toupees can be styled overnight; natural hair comes and goes with the client. Similarly, a pair of shoes left today with the cobbler can be collected at a later date. A service is often perishable, given and received in real time and in situ. Invariably, service involves a face-to-face encounter between a provider and the user. For example, the service provided in a restaurant by waiting staff is at the customer's table. The chef tends to remain in the kitchen, although few would dispute the importance of the chef's participation in the dining out experience. Indeed, the chef's contribution is a vital component of this experience, encompassing as it does menu design, selection and purchase of ingredients, food preparation and cooking, and (immediately before serving) arrangement of the food on the plate. Thus, while absent from the immediate vicinity of the customer, the chef is distanced by time (in earlier parts of the process) and space (when the customer is *à table*). An exception occurs, for example, in a Cordon Bleu restaurant where the chef works in the restaurant rather than the kitchen and cooks dishes at the diner's table. Somewhat predictably, the closer proximity of the chef to the diner (and the absence of waiting staff) reduces the time and space between the person responsible for preparation and the customer. In so doing, the restaurant raises the intimacy levels of personal service. Invariably, a higher price reflects the increased level of culinary expertise, the atmosphere of exclusivity, and the perceived higher quality of the service experience.

Fundamentals of a Service Encounter

At a fundamental level service comprises an encounter between the service provider and a customer. While the customer traditionally encounters a human service provider, there are other features of even the most basic service operations system. The fundamental components of service are shown in Figure 1.1. The figure shows that a service encounter has three essential components: a customer, a service contact component, and a noncontact component.

The contact component is likely to involve person-to-person interaction, although increasingly customer-service provider contact may be through electronic media. Elements of a noncontact component are usually concealed from the customer's sight, although there are exceptions. For example, in a restaurant it's possible for a customer to observe food preparation. This feature is common to both the fast food and fine dining experiences. In a fast food (especially takeaway) eatery, the customer orders from a displayed menu and may select food items already in the cooking process across the counter. In this service situation the preparer and the service provider may be the same person. In fast food places, operating costs may be reduced by multi-skilled employees and further offset by a high throughput of customers. In a fine dining experience, food preparation may be tableside, and thus highly personalized, or the diner may be able to observe the chefs working in an open-plan kitchen. Personalized service tends to be more expensive than mass-produced service and reflects higher levels of culinary expertise and closer attention to personal needs. When ordering bespoke (custom-made) tailoring, the customer may meet the tailor who takes the measurements as well as the front office person who notes the measurements. Customers with sufficient wealth to buy haute couture may meet the designer for personal discussions on the proposed design of the gown. Again, price reflects

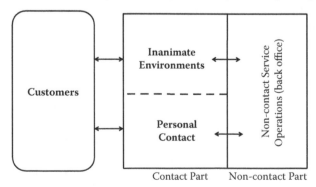

FIGURE 1.1
The service encounter. (From Mitchell M. Tseng, Ma Qinhai, and Chuan-Jun Su (1999), Mapping Customers' Service Experience for Operations Improvement, *Business Process Management*, 5(1), 51.)

levels of expertise, attention to detail, and a designer's reputation. At the low to medium price range, hairdressing is usually open plan, with the stylist working in close proximity with the person washing the customer's hair and other customers. A higher price entitles the customer to increased levels of privacy, personal attention, and exclusivity. When one customer questioned the high price of her hairstyle, allegedly by Vidal Sassoon, he responded the price reflects twenty years of training and hard work.

As Figure 1.1 shows, not only does the customer experience personal contact at the point of service delivery, but there are inanimate environments where the service encounter takes place. These include the ambiance of the location and features of this environment, such as lighting, music, and décor (or not). The value (perceived benefits) that the customer gains from the service encounter is a function of an amalgam of the personal style of the service provider and the inanimate environment that provides the ambiance for the service encounter. The customer takes away the experience, but the inanimate environments tend to stay. Adding to the complexity of service provision may be the presence of other people. Other service users, potential users, and people who have already experienced the service (for example, hotel guests who have been processed through check-in and are waiting for their luggage) may influence service delivery without being a central part in its delivery. When service is located in a crowded place, background noise (chatter, background music, footsteps) may alter and influence the tone of the service encounter. The figure also shows that there are nonservice operations (back office support). Although these are part of the infrastructure for service provision, organizations that rely on service delivery as a major part of their business processes and operations should consider including back office employees as mainstream service providers, for example, in training and educating about customers and their needs. The success (or otherwise) of service often depends on the quality of back office work. Timeliness, suitability, and information accuracy are three key features of back office service support. On those occasions when service breaks down, it is highly likely that this is attributable (at least in part) to shortcomings in back office preparedness.

Key Components of Service

When subjected to detailed analysis, the service encounter is more complex than at first appears. Figure 1.2 shows that a service encounter has three major components: a service delivery system, a service task, and a set of service standards. Of particular relevance are the interfaces between each of these key components, which we have labeled A, B, and C. At these overlapping points service becomes more intriguing and there are inherent risks

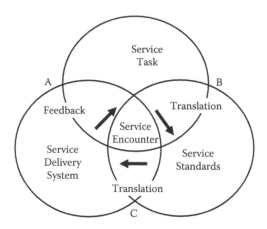

FIGURE 1.2
The service encounter. (From Roger W. Schmenner (1995), *Service Operations Management*, Eaglewood Cliffs, NJ: Prentice Hall International, p. 19.)

for the unprepared or unwary provider of service. It would perhaps not be unrealistic to call these "make or break" points for the organization offering the service. Conversely, organizations that are sincere about their service provision can identify opportunities at these junctures for service differentiation.

Underpinning the service task is the purpose for the service being offered. A service task is within the role and responsibility of the service provider and its employees. For both the service provider and the customer the service task has to convey meaning. Devoid of meaning, the service task has no purpose and loses its edge for both the service provider and the customer. When we asked one interview respondent how she felt about the poor service she'd described receiving from a well-known airline, she said, "I was irritated because their process seemed to make little sense to me." For the customer, meaningless service tasks can be a source of major irritation and are unlikely to lead to repeat use of the service. The phenomenon called customer rage arises in part because of the mismatch between what the customer has been led to expect and the actual experience of the service in execution.[1] Conversely meaningful service tasks are likely to retain the customer for further business.

The roles of the service provider and the customer are in many ways complementary. While the service provider cannot function without a customer, a customer does not necessarily need a service provider, or not specifically *this* service provider. This is especially so when a service is not unique and competing providers offer similar or identical services. In environments where competing outlets are located side by side, a key differentiating feature may be service quality. When the customer and the service provider are one and the same (for example, in self-service environments), the service organization has a number of responsibilities to minimize and obviate uncertainty, confusion, and error on the part of the customer. When customers are expected to deliver the service tasks themselves, the service organization needs to make prior preparations (including pretesting and walk-through exercises) to ensure clarity of instructions to the customer. As an example, witness diners trying to operate the toaster at a buffet breakfast. The absence of a human service provider can make the task of providing quality service more challenging. Delivery via vend-o-mat machines and ATMs (automated teller machines), and service via electronic media such as the Internet places the customer in a dual role: service provider and customer. Customer satisfaction ensues when the technological service provider and the customer work in harmony. When a technological substitute fails to deliver service, and furthermore is mute to the customer's requests, the result is likely to be customer frustration.[2]

Astute managers and supervisors in service environments ensure that there is a human backup available either in person or via a hot line telephone. Although this may seem unnecessary (what might be called a belt-and-braces approach to service delivery), it is likely to forestall customer frustration. In the technology-intensive Tokyo subway system more than a sufficient number of station officials are on duty in the vicinity of the ticketing machines and electronic entry lanes, ready and willing to help passengers use the technology. The vast majority of passengers enter and exit the subway system incident-free, but for the very few who need help, the officials are there. Passenger flow is smooth and efficient, facilitated by competent officials and the highly functioning technology equipment. Rush hours are no exception. When a passenger needs help with tickets from one of the machines, an official is often quick to notice and moves speedily to help. The same cannot always be said for subway systems in other capital or major cities around the world.

Service standards frame the quality of the service task. As such, these can provide a mechanism of control that sets the direction for the service task and within which are parameters for actions by the service provider. Service standards reflect an organization's values; the customer-facing employee is a representative of those values.[3] The relationship of the employee to the employing organization tends to influence the quality of service that the employee is willing or predisposed to provide.[4] Ideally, the list of service standards will be publicized to customers, and customers can have an input into developing service standards (customer feedback forms are one such option).

For customers and the service provider, standards should be transparent and measureable. Service providers need to be aware of how the service provided is assessed from the customers' perspectives. Both the service provider and customers need to have some comparative evaluation of the service being offered against that offered by competitors. It is thus in the interest of service organizations to communicate with, and seek communication from, their customers. A service-based business organization that chose to conduct its operations without recourse to customer interaction and feedback would surely be myopic.[5] But this happens and, as media outlets are not shy to report, within all types of organizations, including multinational corporations (MNCs), enterprises that style themselves as world class.

Conventionally, at the point of delivery to the customer, a service delivery system is face-to-face. Even with the phenomenal growth in service provided electronically, an overwhelming majority of service transactions continue to be up close and personal. In this type of service encounter each participant can respond to the verbal and bodily cues of the other participant. For a number of reasons this renders the encounter more overt and less prone to misunderstandings. When the service delivery is by electronic means there seems to be an increased propensity for error.[6] This is demonstrated by orders for goods and services made by telephone or through an Internet website. Face-to-face service encounters allow participants to notice any possible misunderstandings and to attempt to rectify these in real time. As will be described and discussed in later chapters, customer participation is a critical component of quality service delivery. When a service system has high customer contact, the customer exercises a greater degree of control over the service delivery and may decide timing, content, and quality of the service.[7] For the service organization, the service delivery system is (or should be) the focus of close analysis. If the process of delivery is to achieve acceptable quality standards, the service providers need to make analyses of costs of delivery. Costs of delivery include not only apparent costs such as labor and investment in appropriate equipment (as necessary), but also hidden costs such as staff recruitment and training and maintenance and replacement of equipment over time. Part of the analysis needs to include components to engage customers, especially customer feedback mechanisms. Figure 1.3 shows the symbiotic relationship between the service provider and the service user (the customer). The provision of services from the provider to the customer is complemented by the customer's willingness to provide feedback, a relationship that can be called the service loop.

The three service components of service task, service standards, and delivery system need to be managed so that the three components facilitate the organization to offer service appreciated by customers. The end result of service

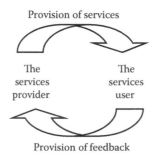

Provision of services

The services provider

The services user

Provision of feedback

FIGURE 1.3
The service loop: symbiosis between the service provider and the service user.

is a collusion of production of service (a back office function that involves preparation for service delivery, educating employees, and fine-tuning the service offering), service task (the result of employee education and practice), and service standards (knowledge and education combined).

As mentioned, we have labeled Figure 1.2 with the letters A, B, and C to designate where each of the key components overlap. Segment A (service task and service delivery system as a contribution to the service encounter, which Schmenner (1995) designates as feedback) is where the service provider may use pre-prepared components of the delivery system. As an example, in a conventional hotel the guests check in via a receptionist who has several roles: welcoming, registering (including the now compulsory credit card swipe), explaining (meal times, times at which facilities open and close, checkout times), and giving directions (to the guest room, to amenities). In some hotels the receptionist may be called upon to escort the guest to the guest room. Often the receptionist is the go-to person for local knowledge and advice, thereby acting in the role of a quasi-concierge. Increasingly, a hotel receptionist is also a salesperson offering a room upgrade (for a stipulated special offer fee), and perhaps at some time during the guest's stay offering loyalty schemes and maybe a time share resort operated by the hotel chain. A new business model for hotels requires guests to manage their own check-in routine, swiping a credit card to obtain entry to the room.

In a conventional hotel, while check-in routines are taking place, the guest can usually presume that the room is ready for occupancy (i.e., has been cleaned, has fresh bed linen, has requisite toiletries in the bathroom, and the refrigerator has been restocked). Sometimes, in a moderately priced hotel at an especially busy time of day some back office tasks (cleaning and preparing the guest room) may lag behind the front office tasks (reception routines). When this happens, a guest can have quite a shock to open the door of a hotel room that has not been cleared and cleaned after the previous occupant. According to hotel workers we have spoken to, when they check out a majority of guests leave the hotel room in an untidy state (bed linen on the floor, all lights switched on, dirty water marks around the bath and shower basin, all towels used and left on the floor).

In a hotel, the range of back office services provides the purpose for the receptionist's service task and tends to be the initial step in the service encounter. Without a customer's need for service, the service encounter is redundant. And, as mentioned, the hotel guest will often be oblivious to this preparation for the stay. Cleaning services tend to be organized and executed out of sync with the main traffic flows of incoming or vacating guests. As with many service tasks, only when matters go awry does a customer become aware of deficiencies in back office delivery systems.

In common with many aspects of service delivery the service user only becomes aware of the service delivery system at the point of the service task. This feature of service may bring problems for service management, as at the time of the service it may be too late to rectify any unforeseen errors in the service delivery process. As an example from a different service environment, the soufflé may be perfect when the chef removes it from the oven (service delivery system), but the dish may be less than perfect when it arrives at the diner's table (an underperformed service task). As with service based around perishable goods, delivery time is critical. In this instance, the service provider's service standards may need review and adjustment, for example, carrying the soufflés, ice creams, and sorbets to the customer's table and returning immediately to the kitchen to collect the steak for delivery to another diner. In service-oriented businesses a key management task is to maintain a smooth flow of the contributing components of service. The customer has a right to expect that these different stages that contribute to the service encounter aim for a seamless flow of service components to be brought together in a timely way at the point of service delivery. A customer's positive experience of service flow contributes to customer satisfaction with a service.[8]

Similarly, the customer has a right to expect that at other stages in the service encounter (such as service delivery and service standards) the service provider is cognizant of health and safety requirements. Thus, the ingredients used by the chef to prepare food are fresh and safe for consumption. And at the cosmetic surgery, the skin medications are not harmful. On a more serious note, the service delivery tasks that contribute to air travel (such as aircraft maintenance, pilot training, pilot shift patterns, and regular pilot health checks) need to receive greater levels of management attention than training in courtesies for the more obvious marketing roles of cabin crew training. From these examples, it is thus apparent that the three key components of a service encounter (service delivery system, service task, and service standards) need to operate in unison. In service, inputs (preparation) and throughputs (especially timeliness) can be as important as outputs.

Segment B (the interface of service task and service standards, which Schmenner (1995) designates as translation) is where the service provider engages the service standards. Crucially this may be where the customer first notices a shortfall in service quality. Poorly trained or lackadaisical employees may be ignorant or indifferent to their organization's standards of service delivery. In the hotel check-in scenario described above, the lack of standards in room preparation may not be evident until the guest opens the door to the hotel room. And then, the lack of quality (whether through oversight, carelessness, or poor record keeping) in this back office service task may tarnish a guest's favorable impression of the receptionist's work (a critical front office task). Where customer service experiences are a mix of good and poor quality, their perceptions tend to reflect the poor rather than the good experience.[9] Inconsistent parts of service delivery thus jeopardize the whole experience from the customer's point of view. Where a service task has several stages, it is not uncommon for the handover between the different tasks to be less than smooth. When the service task falls to the responsibility of several professionals from different parts of an industry, standards may not be comparable. The hairstylist may be a competent professional, but the person who washes and shampoos the customer's hair may forget to test the temperature of the water. In a hospital, the nursing staff may take initial data (height, weight, blood pressure, temperature) swiftly and efficiently, but on entering the consulting room you notice that the doctor is having a bad day, possibly at the end of a long shift.

Segment C (where service standards overlap with components of the service delivery system, and which Schmenner (1995) designates as translation) allows the service-providing organization to audit its service performance. This is made easier if the customer is allowed to provide feedback (and if this is factored into adjusting the service delivery system). In the absence of customer feedback, the service provider operates with incomplete information to aid adjusting the system. Service will achieve quality when the three key components of a service encounter function in harmony, including the fine detail at the interfaces between the three components.

Service and Service Management

In a classic book, *Service Management*,[10] Richard Normann (1943–2003) provides theoretical and detailed descriptions of service and service management and also defines a number of conceptual frameworks. He also asks his readers an intriguing question: Is the management of services different from manufacturing management? By way of explanation he suggests that a number of typical attributes differentiate services from manufacturing. For example, manufacturing production and consumption are separate activities, often separated by space and time. But notice in the Schmenner (1995) model of a service encounter that the service delivery component of service could contain a manufactured element that is later passed to the customer as a part of the service encounter. Manufacturing facilities such as factories and workshops do not usually need to be located near consumers and have tended to be located in out-of-town sites such as industrial estates and industrial parks. Manufacturers may be enticed by local or national incentives to set up their operations in these industrial zones. Health and safety and fire regulations may force manufacturers to locate in these purpose-built areas. There may also be cost gains through lower land prices and cheaper rentals. Local labor markets may be favorable. Industrial premises may be some distance from more urbanized areas where sales outlets are located. In recent decades, out-of-town shopping malls have brought in closer proximity shops and manufacturing, where activities of the latter have not been outsourced overseas.

In a service environment the time between production and consumption is condensed to a point where these activities are near simultaneous. Grönroos (2002) offers a model of the overlapping nature of production

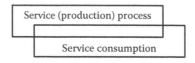

FIGURE 1.4
Overlapping components of service (production and consumption). (From Christian Grönroos (2002), *Service Management and Marketing: A Customer Relationship Management Approach* (2nd ed.), Chichester, UK: John Wiley & Sons, p. 52.)

for service (what Schmenner (1995) refers to as a service delivery system). We reproduce Grönroos's model in Figure 1.4.

In this service environment, the customer (service end user) does not necessarily see the process of manufacture (the service delivery) of the product. More often than not, the customer may only see the finished product when it is offered as part of the service. There are exceptions from a range of industries: bespoke (custom-made) tailoring, dentistry, cosmetic surgery, hairdressing, and beauty and spa treatments all involve the customer in one or more parts of the manufacturing of component parts of the service (creating the service delivery). In a conventional manufacturing process, where a product is manufactured for subsequent sale, the manufacturer has the opportunity between product manufacture and product sales to expend resources on marketing to make the product more attractive to the consumer. Attractive, eye-catching packaging tends to be part of a marketing effort.[11]

Service products tend to be intangible, while manufactured products are more likely to have a concrete form. The innate concreteness of a manufactured product means that it can be protected by various forms of intellectual property legislation. Supporting documentary evidence such as blueprints, materials and design specifications, drawings and photographs, and scale models can show ongoing product innovations. The intangible nature of service means that service innovations can be less easily protected from imitation and copying. This is a double-edged sword, as the uniqueness of a service encounter precludes imitation and copying to a certain degree. To a large extent, service is a social and psychological phenomenon. Service provision takes place through interaction between people. At the least, this is between the service provider and the customer, although others may contribute earlier or subsequent parts of the service. This sociopsychological dimension gives additional challenges in managing service. Service mediated through technology (such as online) challenges the sociopsychological dimension.

Here we revisit and reconfigure a model presented earlier (Figure 1.1). Figure 1.5 shows the revised model when the service employee uses technology to deliver customer service. At its most fundamental, this technology could be a bar code scanner or a machine to swipe a credit card.

The revised model now shows that customer service (contact part) includes personal contact (the employee), inanimate environments such as location, and the technology-delivered service.

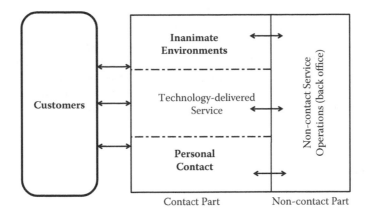

Contact Part   Non-contact Part

FIGURE 1.5
Service when technology supports the employee. (Adapted from Mitchell M. Tseng, Ma Qinhai, and Chuan-Jun Su (1999), Mapping Customers' Service Experience for Operations Improvement, *Business Process Management*, 5(1), 51.)

Transfer of Ownership

Normann also contends that the sale in a service transaction does not necessarily include a transfer of ownership to the same extent as the sale of manufactured products. While we usually concur with Richard Normann, we suggest that there are exceptions to his contention. For example, in a self-service environment the consumer assumes roles conventionally carried out by the service provider, such as making purchase choices, comparing prices of competing products, and physically transporting potential purchases around the shop floor and (eventually) to the checkout. In such a service environment the primary aim of the checkout system is the point of sale (and payment). This is the point at which ownership of the merchandise is formally (legally) transferred from the seller to the purchaser. In self-service environments, the act of transferring ownership has become one of the roles of the purchaser. For example, in a supermarket, customers handle goods before they become their legal property. Some time may elapse between selecting the merchandise and presenting it for payment. In a shop selling goods that are fragile, the sign "Lovely to look at, delightful to hold, but if you break it consider it sold" reminds customers that they have ownership and care responsibilities for the merchandise before actually exchanging money for payment. If we analyze the process of online purchasing of goods or making seat reservations (e.g., on an airline or at a theater), the transaction includes tasks previously undertaken by employees of the seller. Included here can be product recommendations, identification and comparison of product features, price comparisons, selection between product alternatives, and accumulation of several possible purchases before final selection.

Various online businesses now fulfill some of the former roles and tasks of the service organization employee. In the online version of the transaction (toward the conclusion) the purchaser signifies acceptance of the transaction by entering credit card details into the electronic process. The technology then takes over and processes the proffered means of payment. The transaction is completed with acceptance of the customer's credit card details and ownership of the merchandise is transferred to the purchaser.

Service Requires an Immediate User

As mentioned, a service user generally needs to be present at the point of delivery. This is the situation with services as diverse as bespoke tailoring, hairdressing, and spa and medical treatments. In such services, the user is present *because* the service is so personal. Admittedly, one's body measurements can be sent by post or by e-mail. However, this means forgoing the tailor's experience and expert judgment, part of which is sensory (e.g., visual and tactile), which is one key element of a bespoke tailoring service. Once the personal service has been conducted and completed, what has been produced is portable and the customer takes it away. This varies in degree to the type of service. Custom-made tailoring takes time, usually weeks, sometimes months. A manicure or hairstyling takes minutes or an hour at most. The services of a psychiatrist, physician, or dentist may last for a large part of a lifetime.[12] Services from medical professionals may involve a series of repeat visits during which different services are carried out. In using personal services such as health and beauty treatments, a manicure, and hairdressing, customers (usually) go away more contented than before they arrived, though there are known to be exceptions. Noticeably, the proximity of an immediate user is not required with the provision of online services such as travel services, flight and hotel reservations, or Internet banking. This so-called disintermediation of the service from the customer tends to incorporate technology and will be discussed in later chapters.

Service Processes Differ from Manufacturing Processes

It is helpful to contrast the provision of services with the manufacture of products. This brings out differences—both apparent and opaque. A product is manufactured, usually in a specialized environment designed for the purpose, such as a factory, workshop, or design studio. A manufactured product can be delivered, taken away, stored, and used at a later date. However, perishable products (e.g., some foodstuffs in certain types of packaging) are manufactured with a publicized "sell by" date. An exception would be takeaway fast food, which tends to be produced while you wait, although the ingredients will have a sell-by date. Other perishable products may be less obvious. These include airline seats, hotel rooms, and concert tickets. Such products become obsolete after the stipulated date and time. There are very few recorded instances of passengers joining flights after takeoff. Yesterday's hotel room, if unoccupied, cannot be resold today. A ticket for yesterday evening's concert at Carnegie Hall does not retain its currency for tomorrow evening's concert at the same venue. Although the concert may be repeated as one of a series, the booked seat may now be occupied by another concert-goer. In the era when rock stars such as Jimi Hendrix and the Who destroyed their instruments at the end of a performance, the musicians themselves ensured that at the next concert the musician played a new instrument with perhaps a slightly different sound. An exception is probably the Frank Sinatra farewell tour, which lasted several decades.[13]

Manufacture and Delivery of Products Are Separated by Time and Space

Generally, in the manufacture of products, delivery and use are often separated—often by both time and space. In the absence of a waiting customer the manufacturing process has no immediate urgency. Just-in-time (JIT) manufacturing processes are made with the expectation that delivery arrives *just in time* to be incorporated into a prescheduled manufacturing sequence. Where a manufactured item is part of a service, this is not inevitably so. A user does not necessarily need to be present at the point of delivery of the product. But the product needs to be ready for inclusion by the service provider into the service encounter. The book you are holding was written over the past four years in various parts of Europe, Asia, and Australia. Published and printed by an American publishing house, it was delivered as part of a global distribution channel. We do not recall any potential readers of our book being present in our offices, study rooms, or next to us on airplanes as we composed the various chapters—though that would have added an interesting dimension to our creative processes. (Customer feedback usually contributes to a higher quality of service delivery.[14]) However, we conversed with potential readers in our own workplaces and in public spaces such as airports, railways stations, and internationally renowned coffee shops.

Whereas a service user may walk away with the service when provided (dental treatment, a pedicure), the ultimate user of a manufactured product may not necessarily own the product that he or she buys. Examples are the business models of, for example, products (such as some computer software) that are leased to the purchaser/user and not owned outright. Here the purchaser buys the right to use, not the right to own. This also applies when staying in a hotel room for one or more days or nights. In this case, the purchaser buys the use of the room space and amenities for a stipulated time period. It is expected that the room furnishings (television, bed linens, towels, and the unconsumed contents of the refrigerator) remain in the room for use by subsequent guests. Some hotels publish price lists to prevent guests from taking "souvenirs" of their stay. Similarly, restaurant owners and managers expect diners to leave the cutlery and crockery once a customer has used these to consume the meal. The manufactured part of the service provision is the food and the environmental experience, not the utensils provided for the consumption of the meal. Besides, the restaurant makes revenues from preparing and manufacturing the consumables. The equipment used in food preparation and cooking and by the customer is a necessary but not sufficient part of the restaurant experience.

Service Provision Is Personal and Immediate

As discussed, the provision and delivery of a service is personal and immediate. This here-and-now nature gives a psychological dimension to the provision of service. Both the provider and the receiver of the service experience a personal dimension, and each contributes to the personal nature of the service experience. Service is about people: it is up close and personal. The two or more individual human inputs to the service encounter conjoin to create a symbiotic social interaction. Indeed, for a majority of people, experiencing service and engaging with service providers are part of personal social interaction. Often, the collected individual social interactions form part of a greater social intercourse for a community or society. In any part of the world, market days tend to be a big event, often attended by people who travel from afar for the purpose. By definition, a market is where individuals interact to buy and sell. As we recall from studying economic theory, a standard textbook definition of a market is "a place where buyers and sellers meet." The liveliness, and indeed success, of a market is a factor of the interactive transactions between people bringing goods and services to sell and people wishing to experience and perhaps buy. Some markets (such as those specializing in antiques) allot time when the sellers can browse and buy from other traders, often at a professional dealer's discount. A market with many buyers and sellers is lively and interesting. An auction with buyers competitively bidding for the items on sale is an exciting event. An auction with one buyer is likely to be a flop for buyer and seller alike. The atmosphere is augmented by people being there, for example, those who come merely to observe with no intention to engage in commercial transactions. Not everyone strolling down New York's Fifth Avenue, Singapore's Orchard Road, or London's Oxford Street is there to buy. Even without making a purchase, their presence is appreciated by shop owners who are aware of the retailer's adage "a crowd creates a crowd." This has been known since the bazaars of the Orient and the trading houses of the Silk Road, so much so that the creation of "false crowds" is not unknown.

Service Is Emotional

In part because it often engages sensory perceptions, service is emotional. In some situations service is highly charged. Service has been described as "emotional labour."[15] It is suggested that employees (especially public service personnel) should strive to "manage the heart."[16] Emotion is expended by each party to the service encounter. And for good reason this is said to be the "soft side" of service management.[17] The service provider (ideally) possesses social and personal skills to engage the service user as well as specialized knowledge of the service to be provided. In the social encounter that frames the service provision, the service provider adds passion to the personal skills and service knowledge. The service receiver brings an emotional dimension based on prior expectations of the expected service and reframes these emotions in the light of the ongoing experience of the service encounter with the service provider. In collusion there is an emotional engagement. A corollary to these emotional dimensions is the need to develop high levels of trust between service participants. In the current context, trust means having the confidence that the service provider is sincere and can fulfill (make and subsequently keep) promises made to the customer.[18]

The real-time, face-to-face nature of service provision gives immediacy to the social transaction. Service encounters are dynamic.[19] Consequently, customers need to process rapid incoming information—information content that accumulates over the time taken to conduct the service encounter. Service delivery is in real time. This feature of service tends not to allow customers sufficient time to process all of the sensory data. Customers may therefore construct an incomplete picture of the service, especially its quality. This is not to suggest that customers do not assess quality to some degree: some sensory data such as the service provider's personal appearance and tone of voice and the service location convey very important contextual clues to service quality.[20] But in the series of fleeting instances of the service encounter customers may overlook other vital cues and clues that can provide a holistic picture of the service offered. Customers develop a social antenna to sense quality service. Quality may be readily perceived at the extremes of an emotive spectrum: poor or superior quality. Customers, as social beings, are often able to very quickly distinguish between good and poor quality service. A "couldn't care less" attitude on the part of the service provider may be a readily recognizable sign that the service will be poor quality. Superior quality may be equally easily recognizable by a service provider's attentiveness and concern. Service quality may fall anywhere between these two extremes. For both parties in a service transaction, quality perception is a matter of social rapport and sensitivity.

Service and Trust

The trust element means that delivering service is fragile and can be easily damaged. In some business domains the necessity of a trust relationship between the service provider and the service receiver is relatively self-evident. Businesses as diverse as healthcare, wedding services, undertaking and funeral services, medical care, and training (from driving a car to SCUBA diving) have high levels of trust embedded into the service provision. Customers experiencing service in a fine dining restaurant or enjoying an exclusive hotel consider trust to be part of the cost of the experience. Food that is poorly cooked or rooms that are below the customer's preconceived expectations would be an unwelcome surprise. But it happens. Similarly, customers have a right to expect trust to be part of the experience with financial services and support services for so-called high-ticket items, such as those provided for users of exclusive apartments or luxury automobiles. Trust is a critical component when a service has a high emotional content, such as in medical, health, and beauty treatments. Once satisfied, trust can lead to high levels of customer satisfaction and subsequent positive word-of-mouth recommendations for the service.[21] Encouraged by feelings of trust, satisfied customers tend to become loyal customers who are willing (and often enthusiastic) to repeat their experience.[22] Nor are the benefits of trust unidirectional for the sole benefit of the customer. An environment of trust brings long-term advantages to the organization providing the service. These advantages include opportunities for cross-selling other products and services and consistently loyal customers who can play a role in ensuring service quality.[23] Companies such as Apple, Virgin, Prada, and Singapore Airlines (SIA) thrive with a loyal customer base. As trust and continued customer loyalty are critical components in an organization's success, and perhaps even its survival, frontline employees have a key role of providing the appropriate levels of service. Organizations therefore need to ensure that their customer-facing employees (the service providers as far as the customer is concerned) have the necessary skills to convey trust to the service receiver. In a research interview with a hotel director in France, one of the current authors (BH) asked what was the secret of the hotel's success? The answer was brief, candid, and to the point: *"D'abord, on essaie de pas recruire les cons!"*[24]

It is suggested that educating the customer is a necessary component of building a trust-valued service relationship.[25] A trust-based service relationship plays a key role in ensuring high standards of service quality.[26] And, as mentioned earlier, ideally the service relationship is symbiotic.

Trust, once damaged, is often long-term. In some cases a breach of trust may become a permanent schism. Damaged trust (from the customer's perception) can easily spiral so out of the control of the service provider as to make regaining prior trust unfeasible.[27] For a service provider, regaining the trust of customers takes time, effort, and not small amounts of skill. Highly trained providers of customer service can accomplish this. So too can novice providers of service. Minimally trained personnel tend to bring ingenuousness to the service interaction that customers may find refreshing and somewhat charming, as long as it is in small doses. In this environment, customers may disregard service errors as minor infringements of conventional service behavior. Customers may be less likely to excuse lapses in service quality by semiexperienced service providers. Experience suggests that it is more acceptable to be consistently poor or consistently excellent in service provision. Inconsistency tends to be unforgivable. Where service is inconsistent, it is advisable to make a good first impression.[28] Service experiences, whether good or bad, tend to remain in the customer's mind. Bad experiences have an unfortunate habit of remaining in a customer's memory long after the physical experience has rescinded. Customer satisfaction depends not on how good the service encounter starts or what happens in the middle, but how strong it concludes.[29] However, service providers who have limited experience are less likely to possess the full range of knowledge and personal skills to recover a misstep in the service process. In many respects, service failures are inevitable.[30] Retrieving a customer's confidence after service lapse requires a special skill set that is often lacking in new employees or those with limited training and experience.

Endnotes

1. See, for example, Benjamin Schneider and David E. Bowen (1999), Understanding Customer Delight and Outrage, *MIT Sloan Management Review*, 41(1), 35–45; Joyce Hunter (2006), A Correlational Study of How Airline Customer Service and Consumer Perception of Customer Service Affect the Air Rage Phenomenon, *Journal of Air Transportation*, 11(3), 78–109; Paul Patterson, Janet R. McColl-Kennedy, Amy K. Smith, and Zhi Lu (2009), Customer Rage: Triggers, Tipping Points and Take-Outs, *California Management Review*, 52(1), 6–28; Janet R. McColl-Kennedy, Paul G. Patterson, Amy K. Smith, and Michael K. Brad

(2009), Customer Rage Episodes: Emotions, Expressions and Behaviors, *Journal of Retailing*, 85(2), 222–237; Jiraporn Surachartkumtonkun, Paul G. Patterson, and Janet R. McColl-Kennedy (2013), Customer Rage Back-Story: Linking Needs-Based Cognitive Appraisal to Service Failure Type, *Journal of Retailing*, 89(1), 72–87.

2. See Rhett H. Walker, Margaret Craig-Lees, Robert Hecker, and Heather Francis (2002), Technology-Enabled Service Delivery: An Investigation of Reasons Affecting Customer Adoption and Rejection, *International Journal of Service Industry Management*, 13(1), 91–106; Cheng Wang, Jennifer Harris, and Paul G. Patterson (2012), Customer Choice of Self-Service Technology: The Roles of Situational Influences and Past Experience, *Journal of Service Management*, 23(1), 54–78.

3. See Ronald J. Burke (1999), Managerial Feedback, Organizational Values and Service Quality, *Managing Service Quality*, 9(1), 53–57; Dawn Dobni, J. R. Brent Ritchie, and Wilf Zerbe (2000), Organizational Values: The Inside View of Service Productivity, *Journal of Business Research*, 47, 91–107.

4. See Simon Bell and Bulent Menguc (2002), The Employee-Organizational Relationship, Organizational Citizenship Behaviors, and Superior Service Quality, *Journal of Retailing*, 78(2), 131–146; Brian Hunt and Toni Ivergård Toni (2007), Workplace Climate and Workplace Efficiency: Learning from Performance Measurement in a Public Sector Cadre Organization, *Public Management Review*, 9(1), 27–47.

5. Here we allude to the classic article by marketing guru Theodore "Ted" Levitt (1960), Marketing Myopia, *Harvard Business Review*, 38(4), 45–56.

6. Rhett H. Walker, Margaret Craig-Lees, Robert Hecker, and Heather Francis (2002), Technology-Enabled Service Delivery: An Investigation of Reasons Affecting Customer Adoption and Rejection, *International Journal of Service Industry Management*, 13(1), 91–106.

7. Richard B. Chase (2010), Revisiting 'Where Does the Customer Fit in a Service Operation?' in Paul P. Maglio, Cheryl A. Kieliszewski, and James C. Spohrer (eds.), *Handbook of Service Science*, New York: Springer Science, pp. 11–17.

8. See David Xin Ding, Paul Jen-Hwa Hu, Rohit Verma, and Don G. Wardell (2010), The Impact of Service System Design and Flow Experience on Customer Satisfaction in Online Financial Services, *Journal of Service Research*, 13(1), 96–110.

9. David E. Hansen and Peter J. Danaher (1999), Inconsistent Performance during the Service Encounter: What's a Good Start Worth? *Journal of Services Research*, 1(3), 227–235.

10. Richard Normann (2002), *Service Management: Strategy and Leadership in Service Business* (3rd ed.), Chichester, UK: John Wiley & Sons.

11. See the discussion and quality framework in Martin Löfgren and Lars Witell (2005), Kano's Theory of Attractive Quality and Packaging, *Quality Management Journal*, 12(3), 7–20. Also see a development of this framework in Martin Löfgren and Lars Witell (2008), Two Decades of Using Kano's Theory of Attractive Quality: A Literature Review, *Quality Management Journal*, 15(1), 59–75.

12. Famously, Woody Allen began visiting a psychoanalyst in 1959 when he was in his mid-twenties. Psychotherapy forms the backdrop of many of Woody Allen's films and characters. See, for example, Glen Gabbard (2001), Psychotherapy in Hollywood Cinema, *Australian Psychiatry*, 9(4), 365–369. Also see Foster

Hirsch (2001), *Love, Sex, Death and the Meaning of Life: The Films of Woody Allen*, Cambridge, MA: Da Capo Press; Jerrold R. Brandell (2004), *Celluloid Couches: Psychoanalysis and Psychotherapy in the Movies*, Albany, NY: State University of New York Press.

13. This was good news for Neil Sedaka, who wrote the lyrics to "My Way" (original composers Claude François and Jacques Revaux retain the rights to the music). First recorded in 1968, "My Way" became the farewell anthem for Frank's farewell tour. The farewell tour was so popular that the song had many repeat performances over four decades until the iconic singer's death in 1998.

14. Allan H. Church, Miriam Javitch, and Warner W. Burke (1995), Enhancing Professional Service Quality: Feedback Is the Way to Go, *Managing Service Quality*, 5(3), 29–33; Christopher A. Voss, Aleda V. Roth, Eve D. Rosenzweig, Kate Blackmon, and Richard B. Chase (2004), A Tale of Two Countries' Conservatism, Service Quality, and Feedback on Customer Satisfaction, *Journal of Service Research*, 6(3), 212–230.

15. Chih-Wei Hsieh and Mary E. Guy (2009), Performance Outcomes: The Relationship between Managing the 'Heart' and Managing Client Satisfaction, *Review of Public Personnel Administration*, 29(1), 41–57; Brent A. Scott, Christopher M. Barnes, and David T. Wagner (2012), Chameleonic or Consistent? A Multi-Level Investigation of Emotional Labor Variability and Self-Monitoring, *Academy of Management Journal*, 55(4), 905–926.

16. Chih-Wei Hsieh and Mary E. Guy (2009), Performance Outcomes: The Relationship between Managing the 'Heart' and Managing Client Satisfaction, *Review of Public Personnel Administration*, 29(1), 41–57.

17. See discussions in Sriram Desu and Richard B. Church (2010), Designing the Soft Side of Customer Service, *MIT Sloan Management Review*, 52(1), 32–39.

18. Lorraine A. Friend, Carolyn L. Costley, and Charis Brown (2010), Spirals of Distrust versus Spirals of Trust in Retail Customer Service: Consumers as Victims or Allies, *Journal of Services Marketing*, 24(6), 458–467. Also see Deborah Cowles (1997), The Role of Trust in Customer Relationships: Asking the Right Questions, *Management Decision*, 35(4), 273–282.

19. Christian Homburg, Nicole Koschate, and Wayne D. Hoyer (2006), The Role of Cognition and Affect in the Formation of Customer Satisfaction: A Dynamic Perspective, *Journal of Marketing*, 70, 21–31.

20. See, for example, Magnus Sönderland and Claes-Robert Julande (2009), Physical Attractiveness of the Service Worker in the Moment of Truth and Its Effects on Customer Satisfaction, *Journal of Retailing and Consumer Services*, 16, 216–226.

21. Chatura Ranaweera and Jaideep Prabhu (2003), On the Relative Importance of Customer Satisfaction and Trust as Determinants of Customer Retention and Positive Word of Mouth, *Journal of Targeting, Measurement and Analysis for Marketing*, 12(1), 82–89.

22. Simona-Mihaela Trif (2013), The Influence of Overall Satisfaction and Trust on Customer Loyalty, *Management and Marketing for the Knowledge Society*, 8(1), 109–128.

23. Hossein Dadfar, Steffan Brege, and Sedigheh Sarah Ebadzadeh Semnami (2013), Customer Involvement in Service Production, Delivery and Quality: The Challenges and Opportunities, *International Journal of Quality and Service*

Sciences, 5(1), 46–65; Tsung-Chi Liu and Li-Wei Wu (2007), Customer Retention and Cross-Buying in the Banking Industry: An Integration of Service Attributes, *Journal of Financial Marketing*, 12(2), 132–145.

24. A polite translation is: first of all, you try not to recruit idiots!

25. Andreas B. Eisingerich and Simon J. Bell (2008), Customer Education Increases Trust, *MIT Sloan Management Review*, 50(1), 9–11.

26. Shelia Simsarian Webber, Stephanie C. Payne, and Aaron B. Taylor (2012), Personality and Trust Fosters Service Quality, *Journal of Business Psychology*, 27, 193–203.

27. Lorraine A. Friend, Carolyn L. Costley, and Charis Brown (2010), Spirals of Distrust versus Spirals of Trust in Retail Customer Service: Consumers as Victims or Allies, *Journal of Services Marketing*, 24(6), 458–467.

28. David E. Hansen and Peter J. Danaher (1999), Inconsistent Performance during the Service Encounter: What's a Good Start Worth? *Journal of Service Research*, 1(3), 227–236.

29. Richard B. Chase (2004), It's Time to Get to First Principles in Service Design, *Managing Service Quality*, 14(2/3), 126–128.

30. See, for example, Suna La and Beomjoon Choi (2012), The Role of Customer Affection and Trust in Loyalty Rebuilding after Service Failure and Recovery, *Service Industries Journal*, 32(1), 105–125; Shih-Chieh Chuang, Yin-Hui Cheng, Chai-Jung Chang, and Shun-Wen Yang (2012), The Effect of Service Failure Types and Service Recovery on Customer Satisfaction: A Mental Accounting Perspective, *Service Industries Journal*, 32(2), 257–271.

2

Service and Moments of Truth

Service and the Moment of Truth

Service as a social and psychological phenomenon between two or more people is related to the concept of the *moment of truth*, which Richard Normann suggests is critical for the successful provision of service.[1] A metaphor of bullfighting is often used to illustrate the meeting between the provider of a service and the customer. While hopefully service encounters are less confrontational than a bullfight, the sense of engagement (particularly in a symbiotic relationship of mutual dependency) is aptly conveyed. Perceived service quality is realized at this moment of truth. The outcome of this moment depends upon the skill, motivation, and situational creativity of the individual service provider with the customer contributing to the service outcome. A service encounter tends to be outside the control and influence of the service provider's manager or immediate supervisor. In this perspective, it is also difficult and many times even impossible to preprogram the service.

Service as Serial Moments of Truth

The term *moment of truth* refers to encounters between customers and service providers. The moment of truth is that instant when consumers experience and judge service quality—in modern-day parlance, "when the rubber meets the road." This instant (or rather series of instances) often decides service success or failure. Richard Normann writes that the concept of the moment of truth is from bullfighting, where the phrase *"el momento de verdad"* refers to the climax of the spectacle when the matador kills the bull. Since its introduction, many commentators have used the concept of the moment of truth as a metaphor to describe encounters between customers and service providers.[2]

For an organization providing service, the number of moments of truth can run into millions. Jan Carlzon, then chief operating officer at Scandinavian Airlines System (SAS), estimated that in one year of business there are

"50 million moments of truth," and fifteen seconds decide whether the organization succeeds in its service delivery on this particular occasion.[3] Another company identified "152,000 moments of truth each day" or "four customer contacts per second in a typical ten-hour day."[4] It is estimated that, at the very least, a large company in the service sector experiences "tens of thousands of moments of truth a day," and that each of these instances reveals the quality of the service.[5] At each of these moments the service is laid bare and its quality exposed for scrutiny. The consumer has an opportunity to compare the service against expectations and evaluate quality against prior experience with this or a competing service provider. For the customer, these impressions of service are said to be "indelible."[6] Furthermore, the customer may recall other instances of similar service provision (including experiences of competitor's service delivery) and use these experiences to provide a benchmark for purposes of comparison.

Moments of Truth in Practice:
Scandinavian Airlines System (SAS)

Arguably, the first international company to integrate concepts of service management and the moment of truth into its business development was Scandinavian Airlines System (SAS). SAS is a multinational consortium jointly owned by the governments of Denmark, Norway, and Sweden.[7] In 1981, when the airline was losing revenues and faced increased competition for passengers on its routes, the airline appointed a new president and chief executive officer (CEO). The new appointee was Jan Carlzon, formerly president of Linjeflyg, the Swedish national airline. As parts of his turnaround strategy and management philosophy, Carlzon decided that he should focus on communicating his vision and that the airline should focus on its people. Carlzon realized that of the tasks that he and his executives needed to address, a key task would be to develop the SAS human resources (HR) competences. A major challenge within this overall task would be redefining and then inculcating service excellence throughout the whole organization. Jan Carlzon (1941–) became an internationally renowned management guru. Carlzon's book *Moments of Truth* became an international best seller and a must-read book in business and management courses.

The importance of moments of truth gives managers special challenges in a service organization. In particular, the social and psychological characteristics of the moment of truth concept can be particularly challenging. If the service provision is to be successful and effective, an organization's management demands a different type of managerial mind-set. It is not sufficient for an organization to relegate service provision as a responsibility of solely frontline

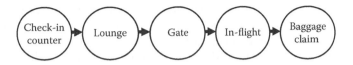

FIGURE 2.1
The travel experience as a service process. (From Fredrik Ekdahl, Anders Gustafsson, and Bo Edvardsson (1999), Customer-Oriented Service Development at SAS, *Managing Service Quality*, 9(6), 405.)

(customer-facing) employees. One of Jan Calzon's executive decisions was to designate all employees to a marketing function; he made the statement: "If you're not working for the customer, you'd better be working for someone who is." This not only conveyed a clear message to employees (such as baggage handlers and maintenance crews) who had never even considered that they were service providers, but publicized throughout the airline's business functions that everyone's work tasks incorporated service delivery.

Figure 2.1 shows a sequence of stages in an airline passenger's journey. Developed by Fredrik Ekdahl and his colleagues for SAS, this is a simplistic model. A walk-through of any airport reveals a multitude of distraction points for passengers (one way that operators of airport terminals generate revenues and profits). Current-day air travel tends to be fragmented with rules, regulations, and procedures hidden from passengers who feel overwhelmed by an invisible, uncaring system.[8]

Figure 2.1 shows the key points at which a passenger interacts with the airline's employees in service encounters during their travel on a particular flight. These key points contain the moments of truth. At each of the five stages in the flight the passenger encounters different SAS employees. For the passenger the perceived quality of service (the delivered value and benefits) is a combination of these moments of truth provided by the different employees to form a connected whole travel experience. For myriad reasons the customer's perceived service quality may be magnificent or deficient, more likely in between these two extremes.

Prior to the arrival of Jan Carlzon, one of the weak points in SAS operations was check-in. Check-in staff regarded their main role as issuing a boarding pass to each passenger booked on a flight. An initial meet-and-greet ("feel good") function was not seen as important as the flight (the physical transportation of passengers from embarkation point to destination). If passengers were to become loyal customers, this clearly had to change. In his strategy to increase passenger revenues, Jan Carlzon introduced business class into the airline's operations. This proposal met with initial resistance from senior management and the airline's union representatives. The rationales were sociocultural: Sweden prides itself on its social equality and frowns upon any hint of social hierarchy or superiority. In terms of a passenger's travel experience, the next stage is the airport lounge. Nowadays, most airlines see a lounge service as a relatively straightforward way to differentiate their service from competitors. Executive lounges have therefore evolved into a somewhat standard service and tend to be publicized by the airlines as a comfort zone where business class and other privileged passengers can enjoy a stress-free environment replete with snacks and drinks free of charge (naturally included in the higher price of the ticket or as part of frequent flier membership).

From the model in Figure 2.1 it is evident that the lounge is the second stage for passengers to interact with SAS employees. Under Jan Carlzon's leadership, the lounges were not only somewhere that passengers interacted with SAS employees, but also where they could access an interactive information system that provided individual passengers with details of their flight, such as ticketing, seating, and other information. A newly introduced SAS kiosk provided this kind of information. Passengers gained access to their data by becoming members of the SAS EuroBonus club (an early example of a frequent flier loyalty scheme). A personalized PIN number used to access the personal flight data ensured passenger confidentiality. SAS kiosks thus gave benefits to both passengers and SAS employees. Passengers who joined the EuroBonus system tended to be people used to being in control of their professional lives and well used to making personal and business-related decisions. The SAS kiosk put such people back in control of their data. In general, as safety concerns are a priority, airports and airlines tend to be rule governed and have become more so after September 11, 2001. For SAS employees, who passengers saw as being able to answer any and all questions about their flight regardless of their job function, the SAS kiosk saved valuable time. The gate, the point of embarkation onto the airplane, is the last occasion that passengers encounter ground-based personnel. At this stage in their travel passengers are obliged to make one of many procedural activities involved in air travel. Procedural activities such as passport checks, airline check-in, and luggage inspections are regulatory and standardized processes that allow and facilitate safe air travel. At the gate the procedural activity involves "exchanging value": surrendering part of the boarding pass in exchange for permission to board the aircraft.[9]

During the flight, moments of truth are many and varied and the airline employees responsible are cabin crew, not ground crew, that the passenger has encountered up until now. On the aircraft, most moments of truth involve human providers of service (such as cabin crew helping stow luggage, demonstrating safety features, and serving meals). If any moments of truth can be said to be up close and personal, it is service provided in the close confines of an aircraft cabin with many passengers requiring service. There are valid reasons why onboard aircraft service tends to be highly structured. First, cabin crew check boarding passes, show passengers to seats and stow luggage, give the safety demonstration, distribute newspapers, hot towels, and cold drinks, and prepare for takeoff. Sequencing, timing, duration, and frequency of services allow cabin crew to maintain control. In this way a series of service activities occupies passengers' attention. A flight is punctuated with technology-enabled service announcements (especially ones involving safety): "please return to your seats," "the seat-belt sign is on," "we'll be experiencing some turbulence." These ensure that passengers are reminded that safety is a concern and that the crew are managing this important aspect of the flight.[10] Emergency announcements are understated: on a recent flight from Bangkok to Hong Kong taken by one of the current authors (BH), an announcement of faulty wiring warranting a return to Bangkok (the point of no return not yet reached) turned out to be an electrical fire. Safely landed at Bangkok some twenty minutes after departure, the large number of fire engines surrounding the aircraft demonstrated that neither the airline nor the terminal operators were taking chances with passenger safety either in the air or on the ground.

Implications of Managing Moments of Truth

In combination, moments of truth reveal features of the organization and its ability to provide service to the customer. Individual moments of truth contribute to the whole perception of an organization's service quality. And an amalgam of individual instances contributes to the perceptions from the market (i.e., customers) toward the organization. It is therefore critical that a service organization successfully manages its frontline service encounters. In the short term customer satisfaction is at stake. In the longer term, the perceived reputation of the organization is at stake. Continued success (or otherwise) of the service encounters will likely determine the survival of the organization.

Scandinavian Airlines Systems (SAS) during the era of Jan Carlzon's presidency (1981–1993) was, as far as we are aware, the first international company to integrate on a large scale the concepts of moments of truth as key foundations of its business growth and strategic development through service management. Under Carlzon's leadership SAS regarded its employees as crucial to business success. Famously, he redefined the business tasks of all employees so that "the entire company—from the executive suite to the most remote check-in terminal—was focused on service."[11] Jan Carlzon was invited to be SAS president when the airline was in a crisis. In the previous two years the airline had posted losses of US$30 million and was rated near the bottom of the European airlines for its lack of punctuality. Within a year of Carlzon's arrival, SAS had returned to profit. By 1984 SAS was voted Air Transport World's "Airline of the Year."[12] During his first years as president, Carlzon initiated 147 projects to improve customer service. With a strategic focus and business emphasis on service, an integral part of the airline's strategic development became development of human resources (HR). In this context HR incorporated development of skills and competences to transform SAS into a service-oriented airline. Three hundred sixty degree evaluations became a yearly process throughout the entire group of companies. In the process Jan Carlzon became a world-renowned management guru (*Moments of Truth* became a best seller). In his introduction to the book, Tom Peters writes: "Carlzon charged the frontline people with 'providing the service they had wanted to provide all along.'"[13]

Daily Moments of Truth

In our everyday lives, we experience countless moments of truth that many of us take for granted. In all likelihood, our subconscious mind will register and remember both the pleasant and unpleasant encounters that we experience with service providers. If we travel as part of our work or leisure routine, we are regularly likely to experience travel and hotel bookings, timetabling, check-in, room service, and checkout. If we are unlucky, we might also experience delayed services, inaccurate bookings, long lines, and mismatched orders. Each encounter may be with a different employee. And, in the case of booking services, the "employee" may be a computer. We experience the different features of restaurant service, such as booking, waiting for a table, the location we are given in the restaurant, face-to-face conversations with the manager and the waiter, the efficiency with which our order is taken, and the speed with which our order is processed and delivered to our table. Also on a day-to-day basis, we experience the services of our mobile phone service provider. The physical instrument for making calls is a minor part of the service and is manufactured by a different organization from that providing the ongoing service. The number of times that we make a phone call is part of the number of moments of truth for the service provider.

If we fall ill, we encounter medical professionals such as nurses, paramedics, physicians, or medical specialists. We might also experience various locations for the delivery of medical services, such as an ambulance, a clinic, or a hospital. Again, each encounter provides a moment of truth between the service provider and the user. In emergencies, speed of service arrival is part of the moment of truth. In medical and similar services, encounters are likely to be laden with emotions such as anxiety, stress, concern, and pain, not only for the service user but also for nearby witnesses, friends, and family members. In such circumstances, managing the moment of truth becomes especially critical. Service in such environments is called emotional labor and requires service providers of a special character and commitment to the customer.[14] Key skills required are the ability to suppress feelings and emotions and maintain a dispassionate demeanor. In this way, the customer and any nearby people do not misinterpret the situation as one requiring panic. Members of the nursing profession, caregivers, and emergency service personnel are rightly renowned for their ability to manage their emotional labor environments.

Treatments such as those offered by a spa or wellness center have their own specialized moment of truth. Such services, situated somewhere between healthcare and the beauty industry, can often relate to a customer's self-esteem by offering a feel-good experience. Even so, trust and emotions may be at a high level, especially in the field of cosmetic surgery, where the service involves a service provider altering a customer's physical features. If the vast majority of moments of truth have fleeting or transient consequences, then those in this area of service may have lasting consequences. And these may be favorable or unfavorable. Cosmetic surgery that enhances the natural contours of the face or body is likely to be regarded by the service user as successful. When cosmetic surgery goes awry, the user is disappointed. If the damage is permanent, the disappointment is worse. This emotion is aggravated when the physician's errors or incompetence is on public view. With a hairdressing failure in cutting, styling, and (especially) dyeing the damage will be rectified as the hair grows. With a cosmetic surgery to the face, this is less possible. Similarly, disappointment with dental work (especially if this involves a tooth extraction) is likely to be more long lasting.

There are myriad travelers' tales of moments of truth at airports. In the golden age of air travel this was the domain of the rich and famous. Lower costs and technological developments brought air travel within reach of a wider range of travelers. As the peripheral services of air travel have become industrialized and mass produced rather than individual, the moment of truth has become less easily manageable for airlines (especially so-called budget airlines). Check-in, baggage handling, in-flight services, and arrivals all bring service encounters to the forefront of a traveler's awareness. Customers' awareness becomes exacerbated as airport procedures and boarding regulations become more stringent and intrusive.[15] For an airline, the industrialization of the flying experience brings special challenges for training, management, and service practices. Where an organization (such as an airline) uses outsourced suppliers to provide customer services, managing these services presents further challenges. In the domain of shopping and retailing, there are moments of truth at each stage of the transaction. From the welcome into the store to the acceptance (or otherwise) of the credit or debit card and the purchase transaction, the service provider has to handle moments of truth. Some of the direct providers of the service will be face-to-face with the customer. Others, such as the financial services provider, will be off-site. However, each provider of service is an integral component of the shopping experience. The success with which the services are provided contributes to that experience. When the service is a pure service encounter, the customer takes away an experience. Where the service is combined with a product, part of the moment of truth will happen when the customer begins using the product. For service providers who deliver pure service, moments of truth tend to happen in real time. Any errors can be adjusted and amended while the customer is engaging in the service encounter. For providers of service that incorporates a product, there is a delayed moment of truth. Away from the location of the service provision the customer is less able to ask the service provider in person to redress any shortcomings.

Governments too have to manage moments of truth in the provision of public services. And, increasingly, these are provided at a distance or electronically. The emergency services (police, fire, ambulance) are one area where it is critical to be successful in service provision. Increasingly, governments are employing electronic aspects of service such as e-government, e-revenue, and online services for education, employment and training, voting, and opinion seeking. The nature of a technological intermediary between the service provider and the customer makes the service encounter less easily managed.

Moments of Truth, the Customer and the Employee

For the customer, moments of truth carry high stakes as customers "invest a high amount of emotional energy in the outcome."[16] As a corollary, the high emotion invested by customers demands an equivalent high performance by the service provider. The moment of truth concept is helpful in defining roles in services delivery and in delineating a framework of services delivery. However, extending the metaphor allows us to note the wider parameters of service delivery, such as the need for training employees to manage impromptu encounters with consumers. As in the bullfighting metaphor, small changes in the service encounter have a great effect. One twist of the bull's head skewers the matador on the horns. Similarly, a customer can alter the potential service quality, perhaps by making unexpected or unacceptable demands for which the service provider is unprepared. There is also a need to educate senior managers (who may not necessarily meet customers on a regular basis). Internalizing service applies to all employees, not solely those whose daily work routines involve direct contact with customers.

Concepts of the moment of truth similar to those that apply within an organization and its relationship to its employees also apply to external relationships of employees with an organization's customers. In this context, the moments of truth for employees are those moments in the workday "when employees directly experience what the company represents."[17] During an employee's working life with an organization there will be many moments of truth. And, as an employee progresses through an organization's career path, that person will build up a mental model of the organization's collective resource capability (human, physical, or other assets). As the organization's "insiders," employees are best placed to gauge the sincerity of statements made to customers about a range of issues, such as quality, timeliness, commitment, and efficiency.

Recognizing Moments of Truth in Service Delivery

In service delivery, moments of truth are as brief as they are in the bullfighting arena. And as a continuum, moments of truth tend to be elongated as part of the process of the service encounter from beginning to conclusion. Moments of truth can be identified throughout the process. It is suggested that for each consumer, two moments of truth are important. The first gains the consumer's attention, and the second provides the experience of the service benefits.[18] Figure 2.2 shows the progression of the service encounter with two moments of truth.

The first moment of truth takes the customer up to the buying decision (i.e., whether or not to proceed with the service encounter). This decision taken, the second moment of truth focuses on the constituent components of the service (including the customer's perceived value and benefits from pursuing the service encounter to its conclusion). In this model the customer is involved at two stages in the process. Within the first moment of truth, up to the buying decision, the customer encounters the "silent salesman." This may be from word of mouth (WOM) from friends or acquaintances, observed advertising, seeing the service in use by others, or point-of-sale publicity. Once the customer makes a buying decision (to buy or not to buy, that is the question), the second moment of truth involves the service provider. This may be an employee of the service-providing organization or a technology (automated machine or online service). In the case of self-service, this second moment of truth depends on the ease of service and payment. For the service-providing organization, managing moments of truth can offer an audit of service quality. In a model called five steps to service

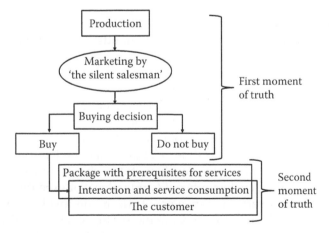

FIGURE 2.2
The first and second moments of truth. (From Martin Löfgren (2005), Winning at the First and Second Moments of Truth: An Exploratory Study, *Managing Service Quality*, 15(1), 109.)

FIGURE 2.3
Five steps to service excellence. (Adapted from Craig Cina (1991), Five Steps to Service Excellence, Building Service Relationships: It's All about Promises, *Journal of Services Marketing*, 2(4), 40.)

excellence, Craig Cina proposes a structured and systematic process of management actions that focus on moments of truth.[19] Figure 2.3 shows the steps in this process. We have adapted the original figure to aid our explanation and discussion.

The bottom of the five steps relates to an organization's awareness of its moments of truth. Organizations should understand that this stage is not wholly straightforward, nor can the exercise be completed quickly. Earlier investigations and analysis emphasize that the task is time-consuming, benefits from a sensitive analysis (including videotaped evidence), and is heavy on human resource effort.[20] Although there may be general patterns, moments of truth are likely to vary from industry to industry. For an organization, knowing moments of truth is the first stage that develops into an inventory of the moments of truth.

Again, care needs to be taken to capture the customer's view of the service encounter. The next step is critical, as it demands assessing the importance of each moment of truth in the service encounter. This is a conjoined assessment of the importance of the moment of truth from the perspective of the service-providing organization and the customer. It is likely that moments of truth gain in importance as the service encounter progresses. The first customer contact with a service employee is an important touch point (Jan Carlzon writes that customer contact with frontline employees lasts an average of fifteen seconds each time). Contacts may be lengthier and more critical as the service encounter develops, takes shape, and seems to have the potential to deliver customer benefit and value. For executives and managers, establishing a service management discipline involves much effort. And, as we show in Figure 2.4, there are increasing levels of difficulty as the organization moves up its process from bottom to top. Implementation, the uppermost step, is predictably the most difficult for executives and managers to achieve success. It therefore makes sense that while each of the five

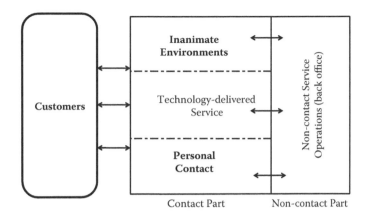

FIGURE 2.4
Technology-delivered service. (Adapted from Mitchell M. Tseng, Ma Qinhai, and Chuan-Jun Su (1999), Mapping Customers' Service Experience for Operations Improvement, *Business Process Management*, 5(1), 51.)

steps is important in alerting an organization's executives, managers, and employees to the moments of truth in their service processes and delivery, at the uppermost steps increased levels of resources need to be mobilized and employed. Not all executives or organizations achieve this uppermost level.

Moments of truth also have relevance in an automated version of service provision. While the majority of moments of truth continue to be face-to-face encounters, encounters facilitated by technology can bypass personal engagement. This feature is shown in Figure 2.4.

The figure shows that by using technology for the required service, the customer bypasses the personal contact and the inanimate environments that surround service delivery. Technology enables the customer to "communicate" directly with the service provider's noncontact service operations (back office). The figure illustrates why using technology to provide service potentially enables organizations to realize economies of scale and reduce operating costs. It is alleged that providing a face-to-face counter service transaction costs a bank around $10; the same transaction machine delivered by an ATM costs the bank around $1, and an online bank transaction costs around 10 cents. Through the use of technology the employee providing face-to-face service is "disintermediated" and is no longer needed to provide service. As the customer can now initiate the service contact, the service encounter becomes more difficult for the organization to manage. The customer can now terminate the service encounter by clicking a computer mouse if, for example, the service is not what is expected. People we have interviewed say they have lost count of the number of times they have clicked to move away from a website in mid-transaction because the booking site was cumbersome to operate. Booking sites for airlines and hotels seem in some cases to be especially user-*un*friendly. When organizations set up a

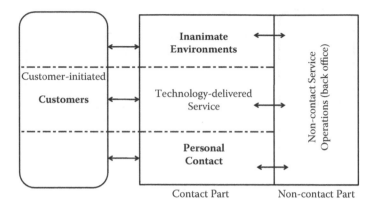

FIGURE 2.5
Service when customers use technology. (Adapted from Mitchell M. Tseng, Ma Qinhai, and Chuan-Jun Su (1999), Mapping Customers' Service Experience for Operations Improvement, *Business Process Management,* 5(1), 51.)

website presence to interface with their customers, this is an opportunity to check and assess whether the quality of their technology-delivered service is acceptable to technology-savvy customers, who may have different quality standards than the organization's usual customer base.

The key message here is: when technology substitutes for a human service provider, the moment of truth needs careful forethought and ongoing management. At the moment of truth the customer has to relate to a technological interlocutor (in essence a machine) instead of meeting a human service provider. This removes the skills of a human interlocutor, who infuses emotion into the service encounter. A human interlocutor can also react to another human as part of the service encounter, including helping to identify and overcome errors and misunderstandings. The moment of truth will fall flat if the customer finds the technology interface difficult to understand, difficult to use, or perceives the technology as unfriendly. A key concept in the design of a technology interface to people is usability.[21]

Endnotes

1. Richard Normann (2002), *Service Management: Strategy and Leadership in Service Business* (3rd ed.), Chichester, UK: John Wiley & Sons.
2. Descriptions and discussions of aspects of moments of truth can be found in Jan Carlzon (1987), *Moments of Truth*, Cambridge, MA: Ballinger Publishing; Gabriel R. Bitran and Johannes Hoech (1990), The Humanization of Service: Respect at the Moment of Truth, *Sloan Management Review*, 31(2), 89–96; Mary Jo Bitner (1995), Building Service Relationships: It's All about Promises, *Journal*

of the Academy of Marketing Science, 23(4), 246–251; William K. Hengen Jr. (1998), Managing Moments of Truth, *Management Review*, 87(8), 56–60; Richard Normann (2002), *Service Management: Strategy and Leadership in Service Business* (3rd ed.), Chichester, UK: John Wiley & Sons; Martin Löfgren (2005), Winning at the First and Second Moments of Truth: An Exploratory Study, *Managing Service Quality*, 15(1), 102–115; Marc Beaujean, Jonathon Davidson, and Stacey Madge (2006), The 'Moment of Truth' in Customer Service, *McKinsey Quarterly*, 1, 63–73; Martin Löfgren and Lars Witell (2008), Two Decades of Using Kano's Theory of Attractive Quality: A Literature Review, *Quality Management Journal*, 15(1), 59–75; Magnus Sönderland and Claes-Robert Julande (2009), Physical Attractiveness of the Service Worker in the Moment of Truth and Its Effects on Customer Satisfaction, *Journal of Retailing and Consumer Services*, 16, 216–226.

3. Jan Carlzon (1987), *Moments of Truth*, Cambridge, MA: Ballinger Publishing.

4. Craig Cina (1990), Company Study: Five Steps to Service Excellence, *Journal of Services Marketing*, 4(2), 39.

5. Richard Normann (2002), *Service Management: Strategy and Leadership in Service Business* (3rd ed.), Chichester, UK: John Wiley & Sons.

6. Mary Jo Bitner, Stephen W. Brown, and Matthew L. Meuter (2000), Technology Infusion in Service Encounters, *Journal of the Academy of Marketing Science*, 28(1), 138–149.

7. Historical background details of SAS can be found in Jan Carlzon (1989), We Used to Fly Airplanes; Now We Fly People, *Business Forum*, 14(3), 6–7; Fredrik Ekdahl, Anders Gustafsson, and Bo Edvardsson (1999), Customer-Oriented Service Development at SAS, *Managing Service Quality*, 9(6), 403–410; Jens Christiansen (2000), *IT and Business: A History of Scandinavian Airlines*, Denmark: Aarhus University Press.

8. See discussion in Fredrik Ekdahl, Anders Gustafsson, and Bo Edvardsson (1999), Customer-Oriented Service Development at SAS, *Managing Service Quality*, 9(6), 403–410.

9. See discussion in Fredrik Ekdahl, Anders Gustafsson, and Bo Edvardsson (1999), Customer-Oriented Service Development at SAS, *Managing Service Quality*, 9(6), 403–410.

10. For a discussion on the dilemma facing cabin crew see Diane L. Damos, Kimberly S. Boyett, and Patt Gibbs (2013), Safety versus Passenger Service: The Flight Attendants' Dilemma, *International Journal of Aviation Psychology*, 23(2), 91–112.

11. Jan Carlzon (1987), *Moments of Truth*, Cambridge, MA: Ballinger Publishing, p. 26.

12. Management guru Tom Peters wrote the introduction to *Moments of Truth* (this quotation is from p. ix). Tom Peters and Robert Waterman were co-authors of *In Search of Excellence* (Warner Books, 1984). At the time of writing this classic management book, both authors worked at the consultancy group McKinsey & Company, New York.

13. Introduction to *Moments of Truth*, by Tom Peters (p. x).

14. See discussions in Sandra Hayes and Brian H. Kleiner (2001), The Managed Heart: The Commercialisation of Human Feeling—and Its Dangers, *Management Research News*, 24, 3–4.

15. See, for example, Lawrence F. Cunningham, Clifford E. Young, and Moonkyu Lee (2004), Perceptions of Airline Service Quality: Pre and Post 9/11, *Public Works Management Policy*, 9(10), 10–25. Also see Garrick Blalock, Vrinda Kadiyali,

and Daniel H. Simon (2007), The Impact of Post-9/11 Security Measures on the Demand for Air Travel, *Journal of Law and Economics*, 50(4), 731–755.

16. Marc Beaujean, Jonathon Davidson, and Stacey Madge (2006), The Moment of Truth in Customer Service, *McKinsey Quarterly*, 1, 62.

17. Keith E. Lawrence (2007), Winning at 'Employee Moments of Truth' through HR Products and Services, *Organization Development Journal*, 25(2), 159.

18. Martin Löfgren (2005), Winning at the First and Second Moments of Truth: An Exploratory Study, *Managing Service Quality*, 15(1), 102–115.

19. Craig Cina (1991), Five Steps to Service Excellence, Building Service Relationships: It's All about Promises, *Journal of Services Marketing*, 2(4), 39–47.

20. See, for example, the various investigative processes described in Fredrik Ekdahl, Anders Gustafsson, and Bo Edvardsson (1999), Customer-Oriented Service Development at SAS, *Managing Service Quality*, 9(6), 403–410; Sara Björlin Lidén and Per Skålén (2003), The Effect of Service Guarantees on Service Recovery, *International Journal of Service Industry Management*, 14(1), 36–58; Subnil Babbar and Xenophon Koufteros (2008), The Human Element in Airline Service Quality: Contact Personnel and the Customer, *International Journal of Operations and Production Management*, 28(9), 804–830.

21. Brian Hunt, Patrick Burvall, and Toni Ivergård (2004), Interactive Media for Learning (IML): Assuring Usability in Terms of Learning Content, *Education + Training*, 46 (6–7), 361–369.

3

Service Management, Service Systems, and Service Excellence

The Moment of Truth: *El Momento de Verdad*

Bullfighting provides the origin of the concept of moment of truth: *el momento de verdad*. The lore and spectacle of bullfighting has attracted award-winning writers such as Ernest Hemingway and James Michener.[1] According to Hemingway, the moment of truth is the climax of the bullfight, a finalization of a process of killing the bull. *El momento de verdad* is "the final sword thrust, the actual encounter between the man and the animal."[2] The earlier parts of the spectacle of the *corrida de toros* (bullfight) lead to the moment of truth. Strict rules of bullfighting prescribe the manner in which the *matador* (bullfighter) should kill the bull. By law, the bull should be killed within fifteen minutes—fifteen minutes in which the matador must create moments of truth. Using a *muleta* (a red cloth folded over a pointed wooden stick), the matador encourages the bull to charge. By skilled use of the *muleta* the matador directs the bull to lower its head as a prelude to a charge. In so doing, the bull exposes the arch between its shoulder blades. Stepping close to the bull, the matador reaches over the horns to the exposed area and plunges the sword into the exposed flesh. Bullfight aficionados recognize this as the moment of truth. The matador plunges the sword deep enough to sever the aorta and cause death. In the timeframe of the bullfight the moment of truth is but a few brief instants. The size and speed of the bull leave little scope for hesitation in this life-and-death struggle between man and animal. If the bull raises its head while the matador is reaching over for the kill, the sword is not long enough to reach the aorta and make the kill. The matador's body is then exposed to the horns. The matador's skillful placing of the sword ensures death. A mistake by the matador in timing and placing the sword (for example, by striking bone) may lead to victory for the bull. If the matador is unable to control the movement of the bull, he is likely to be gored. When this happens, the matador can expect hospitalization—at best.

In the bullring, a matador's survival depends on his ability and actions in response to the movements of the bull. The skill is to get the bull to respond to the movements of the matador.

Ideally, the matador uses his innate knowledge of the bull to predict the bull's likely behavior. In the opening phases of a *corrida, picadors* (men mounted on horses) confront the bull. This is a period of *tienta* (testing) during which the matador observes how the bull reacts. A matador's knowledge and skill are developed over years of practice. Bulls bred for fighting are not exposed to bullfighting until they first enter the bullring. For the matador, practice takes place on farms and in training camps. Matadors refine their performance well away from the bullfighting stadium and paying spectators. Trainee matadors fight cows, young bulls with their horns capped, old full-sized bulls, and full-sized young bulls. By law, bulls should be between three and five years old. Bulls younger than this may lack maturity; bulls older than this are no longer sufficiently agile for the tournament. A bull bred for fighting enters the bullring once only. At the end of the spectacle it is killed. When a matador is unable to kill a bull, the slaughter takes place after the event and outside the view of the spectators. A bull that survives its first encounter with a matador is an extremely dangerous animal. Bulls learn quickly and tend to remember their lessons. For this reason, bulls are not trained with the red flag. Remembering from training that there is no man behind the *muleta*, a canny bull would "forever after ignore the cloth and go for the man."[3] Lacking fear, and knowing the rules, the bull would be the winner in this zero-sum contest.

Ensuring Quality at the Point of Service Delivery

In service delivery and service management, moments of truth are fleeting instances when the service provider and the customer interact. At this moment service quality is exposed for scrutiny. Thus, participants in the service encounter can notice and experience key criteria that contribute (or detract from) the management and implementation of service quality. Each moment of truth tends to be unique. It is thus extremely difficult to teach an employee correct and adequate behavior for service encounters. And as these occur in real time, there is a heavy burden of service responsibility on the employee. Training and other preparations for these types of situations demand special methods and techniques. The moment of truth will clearly show where the service provider has received inadequate training or has poor people skills.

Somewhat ironically, some service organizations and their employees dismiss moments of truth as unimportant and troublesome and seem to regard customers as a necessary irritant. At best, this is a cavalier approach to quality management. At worst, this could contribute to an organization's eventual deterioration and demise. As the popular saying by business consultants has it: "If you think your customers are not important, try doing without them for a month or so." As in the metaphor from bullfighting, at the moment of truth the quality of service is revealed and the customer thus has an unprecedented opportunity to experience the service and assess its quality. Indeed, each moment of truth offers the customer an opportunity "to assess, re-assess, or verify a previously held perception of their relationship with the service provider."[4] Over time, the accumulated sum of individual moments of truth contributes to the customer's overall perception of service quality. Noticeably, an integral part of business development at SAS became human resources (HR) development and the education of the airline's employees in concepts of service management. Key tasks included empowerment and decentralizing authority so that frontline, customer-facing employees could make decisions affecting their service without needing to defer to their line managers. Throughout SAS, 360° evaluations became a standard process on an annual basis for employees at all levels and in all functions.

Richard Normann's (2002) three-phase model of service delivery offers a useful framework for analyzing service.[5] In using and developing this framework, we are able to focus on the tasks for an organization's leaders, managers, and employees. Such an analysis can note in particular the implications for delivering service excellence. In Normann's models of service it is essential to set an organizational-wide framework for the management of good service. The organization as a whole bears the responsibility for service delivery with an essential need to educate all employees, especially (but not exclusively) employees at the forefront of service delivery to the customer. Most important is the role of management in organizational governance. A fundamental need is that senior management must internalize the concept of good service. With this in place, all else follows; in its absence, nothing follows. This is equally important in public and private sector organizations. From some perspectives it is possible to argue that it is even more important in the public sector. However, for public sector organizations the existence of regulatory and governance frameworks increases the complexity of service design and delivery. The need for political control, supervision, and steering brings a more complex set of goals to which cognizance must be paid.

The immediacy of the service delivery (an encounter in real time, often face-to-face) means that it is essential for quality to be "right first time, every time." The rationale for this is twofold. First, there are new customers (whom the organization wishes to retain). Second, there are returning customers (who have prior experience of the service). The personal nature of the service delivery transaction means that effective personal skills are of critical importance and that a key priority for organizations is to educate employees in the required personal skills. This is critical, though often overlooked. The personal component of service provision cannot be overemphasized. Nor can the importance of treating service as part of social

Designing Service Excellence

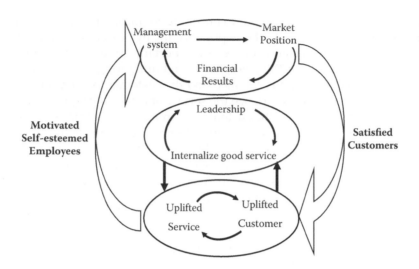

FIGURE 3.1

Components of a service management system. (Adapted from Richard Normann (2002), *Service Management: Strategy and Leadership in Service Business* (3rd ed.), Chichester, UK: John Wiley & Sons, pp. 61–74.)

intercourse. A psychological component adds yet further complexity, as provider and receiver need to determine the parameters of the encounter. This indicates a need for special management skills on the part of the provider. Figure 3.1 shows a three-step model of the processes of service delivery as components of a service management system.

We have developed this model from what Richard Normann calls the virtuous circles of the service company.[6] Virtuous circles are the focuses of service organizations in their efforts to mobilize and direct resources (physical, financial, and human capital) toward satisfying customers. In our adapted model we show an interconnectedness of the internal workings of the service organization, including leadership and internalized service, and the results of these activities. Results reflected in the marketplace include customers' purchasing preferences and the strength of a brand as a service provider. The marketplace also includes financial results that indicate an organization's financial health and strength, as shown in its official documents as well as the more psychologically based views of the financial markets. When organizations show positive financial results from their commercial endeavors, they can in turn make investments in capital assets such as business equipment, buildings, and land for expansion and business development. Arguably the most important investment is in human capital, for example, through recruitment of employees with needed skills, training (perhaps reskilling) employees, and investments in technology, especially that which supports the provision of service to customers.

At the heart of Figure 3.1 is the role of leadership, especially in promoting the internalization of good service. In this context internalization applies equally to the organization's leadership and its employees, regardless of whether their work routines require them to interface with customers. In sum, the organization's leaders and executives should be prepared to "walk the talk." When internalization of service is not embraced throughout the whole organization but restricted only to employees "whose job it is," that organization risks a mismatch between strategic, policy, and motivational announcements made by its leaders to the workforce. This is how organizations become "out of kilter" and lose strategic direction and momentum. Leadership credibility is a prerequisite for a healthy organizational culture.[7] Losing credibility is much easier than gaining it, as credibility needs to be earned by actions rather than words alone.

Moments of Truth as an Opportunity for Assessing Service Quality

For the service provider each moment of truth engaged with a customer provides an opportunity for focused feedback. This includes immediate feedback from customers through their willingness to continue the service encounter to eventual use (purchase) of the service. In this way, the customer can review the proffered service quality against expectations and use these to develop intentions for the next steps in this service encounter.

The service organization also has opportunities to review the service delivery (including the performance of the service employee and using a benchmark of preset service standards). The moment of truth can be the point when service organizations become aware of and reconsider parts of the service encounter that malfunction (suggested by the subsequent actions of the customer, which might include so mundane an action as walking away). Even when the customer's actions are positive (i.e., when the encounter is successful from the differing perspectives of each participant), wise organizations would not be complacent. Rather, they would seek to develop the service standards to higher levels of excellence. From feedback, the organization can also assess when it needs to innovate new elements of service. This service delivery environment has been described as "a laboratory where that part of the innovation destined for the client is worked out."[8] As service involves high levels of human interaction, it is difficult to isolate incidents for analysis in a test tube or a petri dish. The difficulty of observing in real time the large quantities of service encounter data required to make accurate assessments of quality led SAS to set up video cameras at key moments of truth in passengers' journeys. The videotapes were later analyzed by a team of professional researchers.[9] These video recordings of passengers progressing on their journey were supplemented by observations in airport terminals. One executive was reluctant to move out of his office into the terminal to observe passenger flow and was informed that his desk would be moved from his office into the terminal. From that viewpoint, he could make his executive decisions about passenger service. Alternatively, he could remain in his office and delegate his assistant to observe passenger flow in the terminal, in which case the assistant would be empowered to make the executive decisions.[10]

In a survey by McKinsey of the U.S. banking sector, high-performing branches used the moment of truth to identify and resolve customers' problems in retail banking.[11] In this particular study, customers' interactions were "emotionally charged" at the moment of truth with bank employees. Dealing with financial services employees tends to be an emotive experience, especially if the discussion centers on a loan or overdraft arrangement. Affective commitments such as trust (of both bank and employee), reciprocity, shared values (with the bank), and rapport (with the service-providing employee) are influencing factors in financial service transactions and as such need to be addressed at the moment of truth.[12] In the McKinsey study, depending on the employee's response, customers had positive or negative emotions. On the positive side were bank employee responses such as being well informed (for example, about the customer's financial history and current needs), troubleshooting (i.e., resolving) the customer's financial problems, and being proactive in processing financial proposals. Somewhat predictably, in banking services the most positive customer response related to an employee offering good financial advice. In the same study, customers' negative emotions about financial services related to staff incompetence and unfriendliness and lack of suitability in proffered advice and services. Organizations that have ongoing customer relationships create extra value for customers.[13] Included in this extra value are benefits such as security, a feeling of control, and reduced risk in the purchase decision.[14]

A Service Delivery System, including Organizational Systems and Processes

Figure 3.2 shows a service delivery system that includes an organization's systems and processes, its leadership, and the role of customers in providing feedback. The figure should be read in a clockwise direction so that systems and processes lead to mobilization of resources that focus on service tasks. Mechanisms to encourage feedback from frontline employees and customers are designed by the organization's leadership (at either

FIGURE 3.2
Organizational processes for service delivery. (Adapted from Richard Normann (2002), *Service Management: Strategy and Leadership in Service Business* (3rd ed.), Chichester, UK: John Wiley & Sons, p. 73.)

the executive or departmental level). Feedback from service encounters can thus be routed to the leadership for service improvements. With service improvements, the organization generates satisfied customers who become loyal and repeat customers.

In Figure 3.2 we show interlocked components in a service delivery process. The figure shows three components of an organization's environment from the perspective of service management. Two components relate to the organization and one component focuses on customers. Components of an organization are the organizational systems and processes and features of the organization's leadership. The customer-focused dimension relates to customer expectations. Organizational systems and processes include the management systems that relate to how the organization communicates at senior levels and its processes of decision making and mobilizing internal resources (capital assets, financing, and people). A critical component is the culture of the organization ("the way we do things around here").[15] Organizational culture is a crucial component in service delivery and needs to be a major focus for executives who aspire to ensure that their organization can deliver service excellence.[16] The customer-focused dimension relates to customer expectations.

For the organization, critical issues are the management systems and processes, sufficient (and appropriate) internal resources, and an organizational culture that is (or which can become) customer-focused (Figure 3.3). For managers, key tasks are communication, development of employee competencies (through training or brought in via recruitment of new employees), and processes that do not contradict or inhibit customer service and quality. Many organizations seem to have difficulty in achieving these fundamentals.

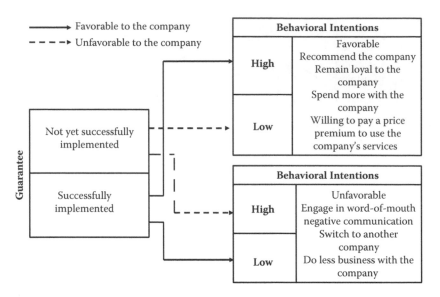

FIGURE 3.3
Effect of service guarantees on customers' behavior (hypothesized model). (From Jay Kandampully and Liam Butler (2001), Service Guarantees: A Strategic Mechanism to Minimise Customers' Perceived Risk in Service Organisations, *Managing Service Quality*, 11(2), 117.)

In mobilizing resources, managers need to address customer issues (such as timely delivery of service) and align competent human and other resources toward this goal. It is important for all within the organization to internalize service and what this means in terms of task performance, team contribution, and consistent quality. Quality of performance is a delicate issue, especially over repeated iterations of service. Technology is capable of delivering quality first time, every time. But technology is reliant on a human element. Human actors tire, need rest breaks, and have bad days. Under these conditions the technology may operate GIGO (garbage in, garbage out). For the organization, the goal is to focus on the service tasks and ensure mechanisms are in place to use feedback for purposes of improvement. Effectively managed, the organization and its processes can expect satisfied customers. The importance and the social and psychological characteristics of moments of truth create a demand for a different type of management to be successful and efficient. Internalization of the good service concept in all parts of the service organization is essential. The management must demonstrate the same high-quality service to all its personnel as the service it expects to be provided to its external customers. Again, this is a goal that many organizations fail to achieve. For Ernest Hemingway, the moment of truth brings satisfaction for the members of the viewing public who pay for seats at the spectacle.

It is important for all personnel in an organization to internalize good service. This includes both back office staff and top management, who may not always meet customers directly. It is said that development of service competences through education and training is wasted effort if training is only provided to frontline (customer-facing) employees.[17] Here, internalizing good service means understanding consumers' needs and how the organization's products and services meet these needs. This can be more successfully achieved with a flat organization. For this reason, in his book *Moments of Truth* Jan Carlzon emphasizes the need for organizations to become flat—especially if these organizations are service-oriented.[18]

Leonard Berry suggests key components of an effective service quality information system.[19] In all there are six features that service organizations should incorporate into a feedback system. Of primary importance are mechanisms that can incorporate customer feedback. Unless this is designed into the system, it will be difficult for the organization to receive and then process valuable feedback from the perspective of customers. A marked distinguishing feature between closed and open systems is that an open system is designed to allow engagement of parts of the system with the external environment, while a closed system is isolated from the environment outside itself.[20] Part of the information system should enable the customer to choose or set priorities for service. Without an awareness of customer priorities executives and managers designing service systems may squander precious resources, including into their service system design features for which customers care little. As a central rationale for focusing on service quality is ongoing improvement to existing service offerings, an ideal information system will process incoming data from the perspective of prioritizing improvements (i.e., where should resources be focused and in which order of urgency). As improvements are made, continuous feedback will aid further managerial decision making on resource preparation, focus, and allocation. Concurrent with service feedback needs to be financial data to plot service improvements against customer revenues. Additional strands of data should relate to the organization's human resources (HR) competences and integrate service improvements with data about employee performance, especially where one or more employees have proposed service quality improvements that generate customers.

Endnotes

1. Ernest Hemingway (2004 [1939]), *Death in the Afternoon*, London: Arrow Books; James A. Michener (2005 [1984]), *Iberia*, New York: Random House Books; James A. Michener (1995), *Miracle in Seville*, New York: Random House Books.
2. Ernest Hemingway (2004 [1939]), *Death in the Afternoon*, London: Arrow Books, p. 59.
3. James A. Michener (2005 [1984]), *Iberia*, New York: Random House Books, p. 20.

4. Kalyani Menon and Aidan O'Connor (2007), Building Customers' Affective Commitment towards Retail Banks: The Role of CRM in Each 'Moment of Truth,' *Journal of Financial Services Marketing*, 12(2), 160.

5. Richard Normann (2002), *Service Management: Strategy and Leadership in Service Business* (3rd ed.), Chichester, UK: John Wiley & Sons.

6. See Richard Normann (2002), *Service Management: Strategy and Leadership in Service Business* (3rd ed.), Chichester, UK: John Wiley & Sons, pp. 61–74.

7. See James Kouzes and Barry Z. Posner (2007), *The Leadership Challenge* (4th ed.), San Francisco: Jossey-Bass; James Kouzes and Barry Z. Posner (2011), *Credibility: How Leaders Gain and Lose It, Why People Demand It* (4th ed.), San Francisco: Jossey-Bass. Also see Brian Leavy (2003), Understanding the Triad of Great Leadership: Context, Conviction and Credibility, *Strategy and Leadership*, 31(1), 56–60.

8. Jean Gadrey and Faïz Gallouj (1998), The Provider-Customer Interface in Business and Professional Services, *Services Industries Journal*, 18(2), 8.

9. See the description in Fredrik Ekdahl, Anders Gustafsson, and Bo Edvardsson (1999), Customer-Oriented Service Development at SAS, *Managing Service Quality*, 9(6), 405ff.

10. See Jan Carlzon (1987), *Moments of Truth*, Cambridge, MA: Ballinger Publishing, pp. 41–85.

11. Marc Beaujean, Jonathon Davidson, and Stacey Madge (2006), The Moment of Truth in Customer Service, *McKinsey Quarterly*, 1, 62–73.

12. Kalyani Menon and Aidan O'Connor (2007), Building Customers' Affective Commitment towards Retail Banks: The Role of CRM in each 'Moment of Truth,' *Journal of Financial Services Marketing*, 12(2), 157–168.

13. Annika Ravald and Christian Grönroos (1996), The Value Concept of Relationship Marketing, *European Journal of Marketing*, 30(2), 19–30.

14. Christian Grönroos (2004), The Relationship Marketing Process: Communication, Interaction, Dialogue, *Journal of Business and Industrial Marketing*, 19(2), 99–113.

15. We quote here from Terrence E. Deal and Allan A. Kennedy (1982, 2000), *Corporate Cultures: The Rites and Rituals of Corporate Life*, Harmondsworth, UK: Penguin Books; reissued by Perseus Books.

16. See discussions in John C. Crotts, Dundan R. Dickson, and Robert C. Ford (2005), Aligning Organizational Processes with Mission: The Case of Service Excellence, *Academy of Management Perspectives*, 19(3), 54–68; Richard S. Lytle and John E. Timmerman (2006), Service Orientation and Performance: An Organizational Perspective, *Journal of Services Marketing*, 20(2), 136–147; Leonard L. Berry and Kent D. Seltman (2008), *Management Lessons from Mayo Clinic: Inside One of the World's Most Admired Organizations*, Boston: McGraw-Hill.

17. See the discussion in K.J. Blois (1992), Carlzon's Moments of Truth: A Critical Appraisal, *International Journal of Service Industry Management*, 3(3), 5–17, especially 11–12.

18. Jan Carlzon (1987), *Moments of Truth*, Cambridge, MA: Ballinger Publishing, especially Chapter 6.

19. Leonard L. Berry (1995), *Leonard L. Berry on Great Service: A Framework for Action*, New York: Free Press, p. 34.

20. Open and closed systems stem from the work of Ludwig von Bertalanffy (1901–1972), who developed ideas that became known as general systems theory (GST). For discussions see Debora Hammond (2002), Exploring the Genealogy of Systems Thinking, *Systems Research and Behavioral Science*, 19(5), 429–439.

4

People and Service: Customers

Service and the Customer

A service organization has two potential types of customer: first-time buyers and repeat customers. The organization has an initial opportunity to impress first-time buyers through the quality of service. If suitably impressed, the customer may become a repeat customer. Repeat customers not only demonstrate loyalty, but also often act as unpaid advocates of the organization's service. Repeat customers, by definition, use a service for a second or subsequent time. They will have prior experience of the service quality of this organization, but perhaps also of the service provided by its competitors. Repeat customers are arguably in a powerful position. First, they have experience and knowledge, and possibly more than some novice employees providing service. Second, they bring high expectations to the service they now are experiencing. They are likely to expect that the current service will be comparable or higher than their earlier experiences of service (including that provided by competitors). This seems a reasonable expectation.

At its core, customer satisfaction rests solely on quality of service provision. Manufactured products can be reverse engineered and ultimately copied. Service quality is arguably the only way in which customers can be really satisfied.[1] In a service encounter, production (creation), consumption, and customer evaluation of a service are concurrent. Thus, frontline employees who provide service need to be adept at identifying consumer needs and expectations, and to be able to do this in real time. Frontline employees also need to be skilled in incorporating these observations into the *ongoing* creative processes of service delivery.

By its very nature service delivery relies on people and their personalities. As an interlocutor in the service process the customer plays a key role in helping construct and shape the service. The customer is an inescapable actor in service delivery and makes inputs to its quality. As any service provider instinctively knows, customers vary. Temperament, age, gender, and nationality are some key differentiating factors. Context is important. Sensitive providers of service look for contextual clues to help them deliver appropriate service that satisfies customer expectations. One of the stated dimensions of customer expectations of service is responsiveness (i.e., responses appropriate to the context).[2] Other dimensions of customer expectations are tangibles, reliability, assurance, and empathy.[3]

Some customers can be impatient, hurried, and edgy (what French hoteliers refer to as *pressé et stressé*). Others may be relaxed and easygoing. Customer expectations may differ. Culture plays a part. People from different cultures seem to have differing concepts of quality and different tolerances of what is acceptable. Some nationalities seem culturally programmed to expect excellence (and to complain vociferously when it is lacking). Others grouse about quality, but under their breath or among their peers. Yet others accept lower levels of quality, perhaps for reasons of personal temperament or because it is culturally inappropriate (bad manners) to criticize and make a scene and complain. For evidence, one only need watch television news footage of passengers in lengthy queues for their airline flights to hear the differing responses.

Customers' perceptions and expectations of service quality differ over time. In general, first-time customers differ from repeat-purchase customers. First-time buyers may have incomplete information of the service and need "guiding through" the process. It is not unknown for first-time buyers to be intimidated, especially if there are many steps in the process or the service is delivered partially or wholly by means of technology. In the UK, when banks first introduced ATMs (automated teller machines), an employee stood outside the bank next to the machine to explain how it should be operated. This has happened throughout the history of technology introduction. The first motor vehicles and railway engines were preceded by a person holding a flag to warn others that the vehicles moved at an astounding speed of several miles per hour. The first person to be killed by a railway locomotive was William Huskisson. The tragedy happened on September 15, 1830, when he fell onto the tracks in front of an oncoming train on the adjacent track. The wheels severed Huskisson's leg, which bled copiously, and he died later the same day. When an organization introduces innovations, a key task for the organization's leader and managers is to educate the customer.[4] Once a particular technology becomes more widespread and more commonplace, it is no longer necessary to use a human to provide warning, instruction, or guidance in use.

It is not easy for service providers to provide consistently high levels of service quality. In informal observations, while we were writing this book, we became particularly sensitive to service and especially service quality. We have now lost count of the number of comments we have made to service providers about their inconsistent service quality. We are also becoming less and less surprised that the service we observe (both to ourselves and others) is inconsistent over time. Our personal experience and the personal experiences of others as reported to us exemplifies how a service can deteriorate over time, and quite quickly.

Reputation Matters

As part of the preparation for this book we have conducted informal interviews with a number of people who are regular users of services. Below we reproduce part of an interview that we conducted with a senior manager in a multinational agency. As a project leader, this person had an official budget to entertain incoming visitors to her organization. She also used her personal funds to treat her team members, usually on completion of a project. From her own professional and social life she had much experience in service. Below is a verbatim record of what she told us about one prestigious restaurant in an internationally famous hotel. The reputation of the hotel and one of its restaurants puts it on the must-do list for tourists and businesspeople alike.

> We used to go to the restaurant there about once a month as a special treat or to celebrate. The location was great. The service was great. Everybody knew what they were doing and the attention to detail was excellent. You could ask any of the waiting staff a question [about the food] and they knew exactly how it was cooked and where the raw materials came from. After they got a new manager, they now push you to buy more. They even charge for the water now. And when you ask a waiter a question about the food, he has to go into the kitchen to ask the chef. And when he comes back to the table you can see he doesn't understand what he's been told. You ask another question and he needs to go off to the kitchen again. So we've stopped going there.

Intrigued by this story and familiar with the long-standing reputation of the hotel and its restaurant, one of the current authors (BH) booked a table to check its accuracy. Inviting a colleague to lunch would be a relatively simple way to confirm details of the story. The restaurant is in an iconic hotel in Bangkok. Rumors had been circulating for some time about a fall in service standards coupled with increases in prices (usually seen as smaller servings on the plate). The view from the restaurant's picture windows is indeed splendid. The restaurant hardware (décor, seating, crockery, cutlery, and tableware) is second to none. And yes, the service experience was more or less the same as told to us. Yes, the waiters pushed the product (suggesting the expensive dishes on the menu and the imported bottled water). Yes, the waiter needed to ask in the kitchen to answer questions about the food. Unfortunately, this is not an isolated case.

Many restaurants on several continents seem nowadays to "push product" and focus on maximizing the profitability per bill. In the hotel industry it is a well-known practice to try and maximize the expenditure (the spend) by each guest. Many top-rated establishments say they prohibit this, and some make a point of insisting that their employees are forbidden to follow this practice. Until fairly recently, maximizing customer spend usually meant informing hotel guests of the hotel's facilities, such as the bar, room service, spa and leisure activities, and special events, such as an Italian night in the restaurant, poolside barbecues, a cultural performance, or live music. Through such events and activities hotel general managers like guests to spend their money in the hotel and try to dissuade them from going outside for their entertainment. Recent years have seen the practice of up-selling become more widespread and extend to restaurant service. For some time, *maître d's* have been aware that a knowledgeable and persuasive sommelier can increase wine sales. Sometimes referred to as the sommelier effect, this relies on customers' needs to consult a perceived expert when they feel that their own knowledge is limited.[9] However, not all persuasive employees are experts in their field.

An additional dimension may come into play where service quality is in no doubt and there is concordance between service task, service standards, and service delivery. In such instances the emotional response of the customer becomes more important.

It has been suggested that only around 4 percent of unhappy customers complain, while the other 96 percent express their dissatisfaction by never returning.[10] So-called customer switching behavior has a negative influence on a company's ongoing operations with a concurrent negative effect on bottom-line profitability. When faced with many choices of venue, diners are not reluctant to try new places, nor are dissatisfied customers slow to give recommendations to their friends and acquaintances of their experiences, whether these are good, bad, or downright ugly. Customer complaints and switching behavior damage a company's reputation, often possibly permanently. As an isolated venue of service provision, cruise ships seem particularly prone to customer complaints about quality. In January 2014 passengers on a Caribbean cruise liner were struck with gastrointestinal illnesses. According to information on the U.S. Centers for Disease Control website (www.cdc.gov), over 20 percent of passengers (622 passengers out of a total of 3,071) succumbed to this type of illness, while of the 1,166 crew members, 50 (4.29 percent) also became ill.[11]

In most large cities, the restaurant industry is extremely competitive. Industry data suggest that 40 percent of restaurants fail within one year of opening for business.[12] Restaurant owners and their employees may misdirect their effort attempting to attract new customers when greater financial rewards can be gained at a lower cost by retaining existing customers. Finding new customers is not easy, especially when an enterprise is new and has no reputation. Building a solid base of loyal customers often is a surefire recipe for success. These two types of business activity have been designated as transactional marketing and relationship marketing.[13] Transactional marketing seeks to attract new customers and tends to focus on elements of service delivered by employees categorized as operations staff. To attract first-time customers, novelty and newness can be emphasized as selling points that may include special introductory offers. The express aim is to bring customers over the threshold to sample the services on offer. Some service organizations repeatedly conduct their marketing so that every announcement and campaign features a special offer. This is short-term corporate thinking. Relationship marketing focuses on the longer term and emphasizes the benefits for the customer by remaining loyal to the service provider. Loyalty programs, frequent user schemes, opportunities to upgrade, and use of executive facilities are examples of relationship marketing, albeit where the service-providing organization buys the customer's loyalty by offsetting profitability for investment in facilities (such as executive lounges). For their part, customers provide personal data in exchange for privileges such as discounts immediately or in the future. By definition, customers need to use the service more frequently (i.e., make more purchases) in order to gain special deals over time.

Transactional marketing is suitable for early stages of customer engagement, such as when an enterprise first opens for business or when services are offered to customers for the first time. Here the need is to publicize the business and its services and "get customers through the door." At this stage, short-term timeframes tend to be more effective than plans for long-term customer involvement. Trial and error can play a part to a certain extent. In relationship marketing, organizations tend to address customer issues for the longer term. Service provision to loyal customers is delicate and is easily broken. Emotional commitment by these customers based on established patterns of trust means that the service-providing organization should take care to educate customer-serving employees in managing different types of customers. This may be overly apparent when customers are differentiated by service standards, for example, by price, location, or membership of an exclusive coterie (such as a loyalty scheme, an "early bird" discount, or frequent user bonus). Often, exclusivity is clearly marked by a border, such as a change in décor in a department store (drapes, furnishings, and floor coverings will be different in the ready-to-wear and *haute couture* departments). Different classes on airlines are indicated by separate check-in desks and boarding priorities landside, and different facilities once passengers are through immigration and airside. For premium passengers there are executive lounges, and on the aircraft differently sized and colored seating and wider spaces, allowing fewer passenger and a less crowded cabin (space in an aircraft cabin is a sore point for passengers). Passengers holding first-class tickets board at their leisure; economy class passengers board by row number. It is said that in first-class air travel one makes friends, in business class one makes acquaintances, and in economy class one makes enemies. Once in the air, at the point of service delivery first-class passengers enjoy greater choice (selection of food and beverage), more space, and cabin crew who are more experienced than their colleagues in the other classes.

Finding first-time customers may be comparatively straightforward. An effective marketing campaign may draw in customers, especially if the venue is new or offers a novel experience. It is comparatively more difficult to retain loyal customers. By definition, repeat customers have experienced the service encounter, have perhaps had opportunities to use alternative service offerings, and have reached a positive evaluation of the alternative service providers.[14] Word-of-mouth recommendations to friends tend to be a powerful incentive to potential customers. Word-of-mouth warnings (caveat emptor) are a powerful deterrent.[15]

The Costs of Poor Service

A useful model and discussion of how poor service impinges on revenues and costs is contained in Zeithaml et al. (1996).[16] Their model is shown in Figure 4.1.

The figure shows four components: service quality, behavioral intentions, customer behavior, and financial consequences. Service quality is subdivided into superior and inferior quality. The customer's behavioral intentions can be favorable or unfavorable with related actions as either choosing to remain a customer or defecting to other service providers. For the service provider the consequences can be positive (+$) or negative (–$). Apart from these direct financial consequences, a service-providing organization may face other consequences. On the positive side, the organization may grow and prosper, in part supported by increased revenues, a customer base that remains stable or grows in size, and customers willing to pay price premiums to enjoy the services. On the negative side, the organization may deteriorate in its perceived quality and over time cease trading. Contributory factors for this may

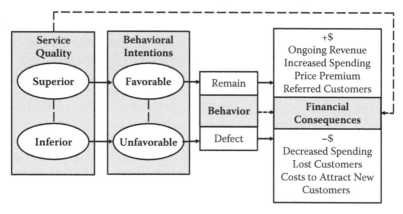

FIGURE 4.1
The behavioral and financial consequences of service quality. (From Valarie A. Zeithaml, Leonard L. Berry, and A. Parasuraman (1996), The Behavioral Consequences of Service Quality, *Journal of Marketing*, 60, 33.)

be a declining customer base attributable to customers defecting to other service providers or increased costs of advertising, marketing, and related publicity to attract new customers. There are also likely to be HR costs (such as for recruiting new employees, including advertising and routines of interviewing applicants) to replace staff. When competent employees join competitor firms, this is not good news. Another consequence of service quality is the enhancement or deterioration of the organization's reputation with its stakeholders (i.e., with customers, employees, and potential customers and employees in the industry).[17] Reputation is also known to affect the share price of listed companies.[18]

As can be seen from Vignettes 4.1 and 4.2, employees, especially those in leadership positions, have a key role to play in attracting and retaining customers. Employees who are proficient in giving good service tend to leave an employer whom they perceive tolerates a poor environment for service quality. As noted earlier, the interfaces between service tasks, service standards, and service delivery need to be in harmony. Where there is discord at these interfaces, the service experienced by the customer is likely to be deficient in some way. We recall enjoying an excellent dinner at a small restaurant in northern Sweden. The *maître d'* (the owner) was very attentive, the food exceptional, the atmosphere warm and friendly. The only thing that marred our full enjoyment of the experience was that our waitress persistently chewed gum, even while speaking with customers and taking orders. The hard-pressed *maître d'* seemed too busy to notice.

Customers tend to avoid organizations where the service encounter is inconsistent or less than expectations. Conversely, organizations that succeed in building a reputation for service excellence tend to have no shortage of applicants wishing to work there and, generally, customers in abundance. Competent employees who provide high-quality service are central to delivery of service to customers. In the eyes of some customers, this type of employee embodies the service encounter. This phenomenon can occur in so-called pure services where the service is a major component of the service encounter.[19]

Loyal Customers

Loyal or repeat-purchase customers have the benefit of prior experience. They may, for example, know the purchase process better than a novice service provider. This adjusts the balance of control in the service transaction. Customers making a repeat purchase may be doing so on the basis of an enjoyable earlier experience: an experience that they not only seek to replicate, but also to better with their next service encounter.

Vignette 4.1

CUSTOMER SERVICE ENCOUNTER

Encouraged by its wide advertising in local media, we tried a new Italian restaurant in Bangkok. As we were some months into writing this book, we constantly sought examples of service excellence. We were not disappointed. The *maître d'* was friendly, attentive, and knowledgeable. Table service was efficient, provided by a courteous waiting team. Throughout the meal service was timely. The various courses were delicious. Unobtrusive background classical music added to our dining pleasure. The décor was modern and tasteful. Someone had obviously invested heavily in this venture. We ordered cocktails. As the restaurant was quiet small, we could see these being prepared. Allowing customers to observe their service order being processed is a key component of a service experience, as it makes them feel part of the experience. Hence, many restaurants now have open kitchens. In terms of the Schmenner's (1995) model, it allows the customer to observe and assess service standards. The food menu arrived with the cocktails. When we appeared ready, the *maître d'* took our order and read it back to us.

Overall, the whole dining experience was memorable. Not only were we delighted that the service encounter satisfied our upbeat expectations, but we also noted with interest that it fulfilled each of the three components of our service encounter: service task, service standards, and apparently the service delivery system. In terms of the task, we felt that our needs were attended to by both the *maître d'* (who welcomed us into the restaurant and showed us to a table) and her waiting and kitchen personnel. We had not made a reservation, as the area contained many restaurants of various cuisines, styles, and prices. We knew the location well, as it was quite close to one of our offices, which we were using as our base of operations for our files and other documents for this book. The *maître d'* offered us the drinks list in a timely manner (not immediately when we sat down, but not so that we were sitting expectantly at an empty table). Clearly, this person knew something akin to the 3-7-11 rule about restaurant service: greet the customer within three minutes and offer to take the order within seven minutes. If eleven minutes elapse and you've ignored your customer, he or she will leave and won't come back.

ANALYSIS OF THE SERVICE ENCOUNTER

We analyzed our service encounter from the perspectives of the three key components proposed by Schmenner (1995): service task, service standards, and service delivery system. As the key person responsible for the service task, the *maître d'* was an exemplar of her profession. In the opening moments of this service encounter we were each struck by her friendly approach and her attention to our requirements (e.g., table location, preprandial cocktails). She was knowledgeable about cocktail preparation and the menu, including that day's specials. We saw that in her supervisory role of the waiting staff she was observant but not overbearing: her staff carried out their tasks naturally without overt instruction. The staff managed to be efficient and pleasant but not a noticeably constant presence near diners' tables. Crowding the customer is an irritant that knows no bounds and seems to be done in some fine dining as well as in mid-range restaurants.

Courteous service was of a high order. Service standards were high, especially considering that the restaurant was relatively small (seating thirty diners at most) and moderately priced. It was obvious to us that service standards were shared by all employees. There were no obvious flaws in the service fabric. In our meal of several courses, each course arrived evenly paced with no apparent rush and no noticeable lassitude. Each course was well prepared and well presented on the plate. We asked for our compliments to be passed on to the chef and kitchen team. Throughout our meal there were other diners, building up over the space of one hour to around 60 percent of dining capacity. We overheard the favorable comments of other diners. As the restaurant was around six weeks from its opening night, it had a winning formula. Clearly it had a potential for a promising future.

While paying the bill we chatted with the *maître d'* as we were keen to learn the secret of this success. Our discussion was most enlightening. She told us that she did not have any family connection to the hospitality or restaurant business but had become interested after watching cookery programs on television. She had worked in restaurants to support her studies and had received a small scholarship to study catering and hospitality at a Midwest college in the United States. While studying she had worked in a local restaurant as a waitress. She had been graduated for less than one year. On paying our bill we took a glance into the kitchen through the swinging doorway. Three people were on duty: a chef and two assistants. We noticed that the kitchen was immaculate, as spotless as the proverbial operating theater.

We left a generous tip. We attributed the success of this service encounter to three factors. First, leadership was in the hands of a trained and experienced professional who clearly had a passion for her chosen craft. As such, she was an excellent role model for the rest of the team. Second, there was a smooth transition between the three key component parts of the service encounter, what Schmenner (1995) labels feedback and translation. As the saying goes, "You couldn't see the join." The three components of task, standards, and delivery were mutually supportive. The service task was executed to obviously high standards (shared by all employees), standards translated into effective food preparation (including taste, presentation, and timeliness), and the delivery system underpinned the service task so that customers were elated. Between ourselves, we discussed the idea of returning to ask permission to interview key employees and to incorporate their practical knowledge and wisdom into this book. With this in mind, we revisited the restaurant three weeks later.

Vignette 4.2

CUSTOMER SERVICE ENCOUNTER

Immediately upon entering we noticed that the restaurant had a different feel. For a start, although this was mid-evening, there were no other customers, no music, and no atmosphere. The ambiance was less friendly. Indeed, it was somewhat unwelcoming. We were not greeted on arrival and made our own way to a table. Waiting staff were standing around. We recognized no one from our earlier visit. As we were to discover, the service experience was also much less efficient. In fact, it was in shambles. We were to experience a living example of Murphy's law, sometimes colloquially (and ironically) referred to as the fourth law of thermodynamics.[5] Murphy's law states: "Anything that can go wrong will go wrong." We had stumbled into a situation where the clueless seemed to be leading the careless, perhaps even the carefree.

We sat down and read through the menu. We noticed that there were breadcrumbs on the table (but no bread basket). As there were no other customers, we supposed that these must have been there since lunchtime. We touched a crumb or two and they were dry and brittle. This was a poor start. We asked a passing waiter to clean the table. The experience became progressively worse. The service was abysmal. The *maître d'* was clueless, the waiting staff clumsy to the point of embarrassment. This *maître d'* (different from our previous one) seemed incompetent even to run a bath let alone a restaurant. We ordered; nothing was read back to us for checking. We asked the waiter to do this. There were three errors in the written order (we'd only ordered four dishes). We asked another server for a drinks menu. We thought it best not to trust anyone to make cocktails. Somewhat against our own better judgment we ordered a bottle of wine. At least the contents would be protected by a cork.

The waiter had difficulty first finding and then using a corkscrew, and while pouring wine into our glasses, spilled some wine on the tablecloth. We noticed that other tablecloths were wine speckled. Our courses arrived out of sequence. Appalled and fascinated in equal measure we began taking notes. We looked around for the hidden TV camera. We half expected an appearance of Basil Fawlty.[6] One of us walked outside and checked the restaurant name and the premises on either side. This was the same place as before, even if the experience was as different as chalk is from cheese.

Throughout the meal we needed constantly to remind the servers to bring bread, to bring condiments, and to top up our glasses of water. Having lost trust in the waiting staff, we kept the wine bottle well away from them and poured the wine ourselves. We didn't spill any. We reminded ourselves that this was not a hole-in-the-wall place, but a moderately expensive place. We eventually got through the meal. At least the kitchen crew knew their job, so the service delivery component functioned well. We declined to order dessert or coffee. We checked and rechecked the bill. We paid. We did not leave a tip. We took a glance into the kitchen; it was still immaculate. No change there it seemed. Again, the whole experience was memorable, but this time for all the wrong reasons.

What was the cause of this service fiasco? A different *maître d'* was on shift. Poor leadership at the senior level had a knock-on effect throughout the whole process and beyond. We recalled a quotation from Jan Calzon's *Moments of Truth*: "The right to make mistakes is not equivalent to the right to be incompetent, especially not as a manager."[7] We asked to speak with the *maître d'*, who had singularly ignored us throughout our meal. We enumerated what had gone wrong, gave our comments as customers, and offered suggestions for improvement. We were listened to politely, but without comment or apology. We had the impression that our customer views were regarded as both unwelcome and an irrelevance. Four pairs of customers who'd arrived after us were receiving the same poor service. We heard them, as we had done, reminding the staff to bring components of their meal (so much for practice makes perfect). We at least did not consider our suggestions irrelevant.

We asked after the earlier *maître d'* and were informed that she'd left to work in another restaurant. No surprises there, then.

ANALYSIS OF THE SERVICE ENCOUNTER

We adjourned to a different restaurant to reflect on our experience. We used the same framework as before to make sense of our service experience.[8] Obviously, this time around it had been a "chapter of accidents." What was the cause of this service fiasco? Noticeably, except in the kitchen, there was a marked change of personnel. Indifference seemed to have replaced professionalism, and laziness seemed to have replaced self-esteem. Leadership was poor (i.e., nonexistent). The interfaces between task, standards, and delivery were shaky at best. As we'd discovered, the link between quality and service delivery (food preparation and cooking) retained its original gloss. A customer seldom sees service delivery and may only notice when things go wrong. In service-intensive industries such as the hospitality industry, the customer experiences the quality of service delivery at some time after the point of contact with the service task. Even so, customers may not experience service delivery in its entirety. Unless their stay is for longer than an average time period, hotel guests may not experience all of the amenities on offer. Lifestyle preferences may preclude some service users from a full range of services: vegetarians do not experience the steak, and carnivores may not opt for the vegetarian meals. Guests who prefer to be less active are likely to avoid the fitness room and the tennis court.

In our second service experience at this restaurant it was fairly obvious that the critical mismatches were between service delivery and service task, and between the service task and service standards. In detail, the readiness of the public space (the restaurant) did not match the readiness of the private space (the kitchen). Ironically, the public space was of a lower order of readiness than the private space, which the customer usually doesn't see. This was evidenced by the quality of food, which suggested that the kitchen team were up to their service task, while the quality of the serving staff (including the *maître d'*) was below standards. The mismatch of quality at the key interfaces between task, standards, and delivery made a noticeable difference. While there seemed to be competent leadership and professional skills behind the kitchen door, such skills were not evident "front of house." We reminded ourselves that at the end of our first visit we'd made plans to return with colleagues and overseas visitors. We'd asked for a number of the restaurant's business cards, and over the following few days, we'd handed these out to friends with glowing recommendations. After the second encounter we advised these same friends that we'd made a mistake. We decided that we'd return to some of our usual venues for hosting dinners for our incoming visitors.

To keep our sense of perspective, we asked a colleague to critique the service in a new restaurant she tried. The restaurant had been open for less than one month.

> The restaurant is difficult to find as it's near the end of a small road off the main road. We'd got a map off the Internet but needed to phone the restaurant to check. After making a U-turn we got there eventually. The restaurant is in a renovated traditional-looking house, and they've made a good job of the interior design. The owners are French and Italian, and they're both very friendly. They greeted and welcomed us. As appetizers we were served Prosecco and olives (the large ones that are delicious). The restaurant was not crowded, but it had a "buzz." The cutlery and crockery were in keeping with the stylish décor. The food was mainly Italian. They had a degustation of around seven courses, but we thought that was a bit too much. The food was cooked well and there was enough on the plate that you didn't feel hungry at the end. Everybody was friendly, but at times a bit overenthusiastic. The prices were reasonable, and at other places you'd get less for the price. We thought this was good value for money. Would we go there again? Yes, for sure.

This critique seemed to address all of the concerns that diners have when going to a restaurant for the first time. The first task is finding the place. While some new openings are accompanied by much hype apart from an Internet presence, this restaurant seemed to take a low profile. Making a phone call to check location and directions is an indication of a customer's time investment in the decision (some first-time customers may have just given up). Once at the venue, the welcome (a moment of truth) is important. For these diners, the restaurant owners seemed to do this well. The ambiance outside and inside set a high standard: the renovated traditional building, the friendly welcome, the large olives, the "buzz." From this beginning the service and the food also needed to be of a comparable high quality. As the critique indicates, this was the case. Food quality (preparation and cooking) and price gave these customers what they considered to be value for money. They expect to turn this first visit into repeat business for the restaurant.

Curasi and Kennedy (2002) identify and categorize five levels of repeat customers and loyal customers.[20] In this classification, detached loyalists have a low commitment to the service provider but are prevented from defecting to other service providers due to high switching costs. Another category, purchased loyalists are locked in to the service provider by loyalty schemes (such as frequent user programs).[21] Designing such programs to lock in customers seems to make business sense, as they prevent customers from defecting to other service providers.

Satisfied loyalists are categorized as reasonably satisfied and (at least at present) have no cause to defect to another service provider. This type of customer has moderate to high levels of commitment, such that while there may be many alternative service providers who can satisfy this customer's needs, there are no current or pressing reasons to switch service providers. Occupying the highest level of the Curasi and Kennedy (2002) typology are apostles. These customers have an emotional attachment to the service providers, and service providers often have an emotional attachment to these customers. Even though there are likely to be many alternative service providers, such customers have high commitment. Commitment needs to be bidirectional: from service provider to customer and from customer to service provider. Apostles tend to be advocates of the service they receive and make positive recommendations to friends.[22]

According to Adrian Payne, customers can attain a higher level of development than advocating the service quality to others. This can be shown as one rung of the loyalty ladder (see Figure 4.2).

The loyalty ladder shows progressive rungs in a customer's relationship with a service-providing organization.[23] This relationship begins when the customer is a prospect approaching the service for the first time. The customer-service provider relationship increases in intensity until the customer is a "partner" who is part of the organization and is actively involved in shaping the service. At this stage the customer contributes to the added value (customer benefits) gained from using the service. Intermediate rungs on the loyalty ladder are as a client, supporter, and advocate. As the customer relationship develops, the organization and its frontline employees need to be aware of the changing parameters of the relationship. Customers who are prospects may perhaps be shopping around, and thus have little initial emotional attachment to the service provider. Customers who become advocates and partners show a greater level of emotional attachment to the service provider. Trust is a key part of this emotional investment, and needs to be reciprocated by the service provider. For example, advocates publicize the service they receive to others, and thereby put their own reputation at stake. We were somewhat premature in recommending to friends and colleagues service that we experienced only once (refer to Vignettes 4.1 and 4.2).

FIGURE 4.2
The customer loyalty ladder. (From Adrian Payne (1994), Relationship Marketing: Making the Customer Count, *Managing Service Quality*, 4(6), 29–31.)

A customer progressing through the rungs of the loyalty ladder should be entitled to expect different levels of service quality. By extension, the customer is also in a position to assess the quality of the organization. Customers who enjoy the service experience can become loyal customers. Some may even become ambassadors and advocates of the organization and publicize the organization to friends and acquaintances, detailing the favorable and unfavorable behavioral responses made by customers in response to promises made by service personnel.[24]

A customer's behavioral intentions toward the relationship with the service provider can be either positive or negative. A model of these behavioral responses is shown in Figure 4.3.

Both types of behavior affect the service-providing organization. Positive behavior patterns generate higher revenues from this type of customer through his or her own purchases and the purchases of others to whom he or she recommends the service. Conversely, customers with a propensity toward negative behaviors reduce their own spending with the service-providing organization. Such customers (if they remain so) also do not recommend others to use the service and may criticize the organization. Over time, reduced numbers of customers threaten the survival of any organization. When service provision is a major component of an organization's business model, then the threat becomes more severe. As the title of a recent book suggests, it's all about service.[25]

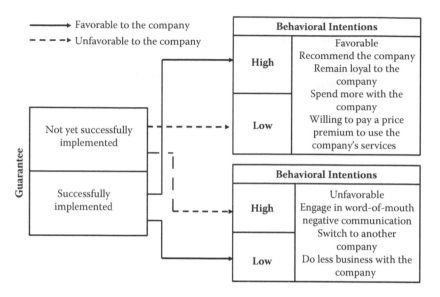

FIGURE 4.3
Effect of service guarantees on customers' behavior patterns (hypothesized model). (From Jay Kandampully and Liam Butler (2001), Service Guarantees: A Strategic Mechanism to Minimise Customers' Perceived Risk in Service Organisations, *Managing Service Quality*, 11(2), 117.)

Endnotes

1. See, for example, discussions in Steven A. Taylor and Thomas L. Baker (1994), An Assessment of the Relationship between Service Quality and Customer Satisfaction in the Formation of Customer's Purchase Intentions, *Journal of Retailing*, 70(2), 163–172; Valerie A. Zeithaml, Leonard L. Berry, and A. Parasuraman (1996), The Behavioral Consequences of Service Quality, *Journal of Marketing*, 60, 31–46; Matthew L. Meuter, Amy L. Ostrom, Robert I. Roundtree, and Mary Jo Bitner (2000), Self-Service Technologies: Understanding Customer Satisfaction with Technology-Based Service Encounters, *Journal of Marketing*, 64, 50–64.
2. A. Parasuraman, Valarie A. Zeithaml, and Leonard L. Berry (1985), A Conceptual Model of Service Quality and Its Implications for Future Research, *Journal of Marketing*, 49, 41–50.
3. A. Parasuraman, Valarie A. Zeithaml, and Leonard L. Berry (1985), A Conceptual Model of Service Quality and Its Implications for Future Research, *Journal of Marketing*, 49, 41–50.
4. Steven H. Stepanek (1980), Educate Your Customers to Appreciate Your Service, *Business Horizons*, 23(4), 21–22.
5. Somewhat predictably, Murphy's law is said to have originated in the U.S. military around the end of WWII. Some histories give Murphy's law an earlier provenance. Fingal's law is a development of Murphy's law and states: "Anything that can go wrong will go wrong"—at the worst possible moment.

6. Basil Fawlty is the main character in the British comedy series *Fawlty Towers* about a hotel where the service is faulty (hence the title of the program). The program was first broadcast in 1975 and quickly became a byword for poor service, especially in hotels and restaurants. John Cleese (of *Monty Python* fame) played Basil Fawlty as a class-conscious officious bore who bullies his staff but who is in turn bullied by his wife. John Cleese also wrote the script for the program together with his then wife Connie Booth (who plays a waitress in the program).

7. Jan Carlzon (1987), *Moments of Truth*, Cambridge, MA: Ballinger Publishing, pp. 83–84.

8. Roger W. Schmenner (1995), *Service Operations Management*, Eaglewood Cliffs, NJ: Prentice Hall International, p. 19.

9. Melissa Manske and Glenn Cordua (2005), Understanding the Sommelier Effect, *International Journal of Contemporary Hospitality Management*, 17(7), 569–576. Also see John Hall, Larry Lockshin, and G. Barry O'Mahony (2001), Exploring the Links between Wine Choice and Dining Occasions: Factors of Influence, *International Journal of Wine Marketing*, 13(1), 36–53.

10. See, for example, Jerry Plymire (1991), Complaints as Opportunities, *Journal of Services Marketing*, 5(1), 61–65; Beatriz Moliner Velázquez, María Fuentes Blasco, Irene Gil Saura, and Gloria Berenguer Contrí (2010), Causes for Complaining Behaviour Intentions: The Moderator Effect of Previous Customer Experience of the Restaurant, *Journal of Services Marketing*, 24(7), 532–545; Jooho Kim and Soyoung Boo (2011), Influencing Factors on Customers' Intention to Complain in a Franchise Restaurant, *Journal of Hospitality Marketing and Management*, 20(2), 217–237.

11. http://www.cdc.gov/nceh/vsp/surv/outbreak/2014/january21_explorer_seas.htm.

12. Ray Arora and Joe Singer (2006), Customer Satisfaction and Value as Drivers of Business Success for Fine Dining Restaurants, *Services Marketing Quarterly*, 28(1), 89–102. See also Ruth N. Bolton and Tina M. Bronkhorst (1995), The Relationship between Customer Complaints to the Firm and Subsequent Exit Behaviour, *Advances in Consumer Research*, 22, 94–100.

13. Adrian Payne (1994), Relationship Marketing: Making the Customer Count, *Managing Service Quality*, 4(6), 29–31.

14. Noel Siu, Tracy Zhang, and Cheuk-Ying Yau (2013), The Roles of Justice and Customer Satisfaction in Customer Retention: A Lesson from Service Recovery, *Journal of Business Ethics*, 114(4), 675–686.

15. Alex M. Susskind (2002), I Told You So! Restaurant Customers' Word-of-Mouth Communication Patterns, *Cornell Hotel and Restaurant Administration Quarterly*, 43(2), 75–85. Also see Ute Walter, Bo Edvardsson, and Åsa Öström (2010), Drivers of Customer Service Experiences: A Study in the Restaurant Industry, *Managing Service Quality*, 20(3), 236–258.

16. Valarie A. Zeithaml, Leonard L. Berry, and A. Parasuraman (1996), The Behavioral Consequences of Service Quality, *Journal of Marketing*, 60, 33.

17. Gary Davies, Rosa Chun, and Michael A. Kamins (2010), Reputation Gaps and the Performance of Service, *Strategic Management Journal*, 31(5), 530–546. See also Shih-Ping Jeng (2008), Effects of Corporate Reputations, Relationships and Competing Suppliers' Marketing Programmes on Customers' Cross-Buying Intentions, *Service Industries Journal*, 28(1), 15–26; Kavita Srivastava and Narendra

K. Sharma (2013), Service Quality, Corporate Brand Image, and Switching Behavior: The Mediating Role of Customer Satisfaction and Repurchase Intention, *Services Marketing Quarterly*, 34(4), 274–291.

18. See, for example, John M. Hannon and George T. Milkovich (1996), The Effect of Human Resource Reputation Signals on Share Prices: An Event Study, *Human Resource Management*, 35(3), 405–424.
19. Michael R. Solomon, Carol Surprenant, John A. Czepiel, and Evelyn Gutman (1985), A Role Theory Perspective on Dyadic Interactions: The Service Encounter, *Journal of Marketing*, 49(1), 99–111.
20. Carolyn Folkman Curasi and Karen Norman Kennedy (2002), From Prisoners to Apostles: A Typology of Repeat Buyers and Loyal Customers in Service Businesses, *Journal of Services Marketing*, 16(4), 322–341.
21. Byron Sharp and Anne Sharp (1997), Loyalty Programs and Their Impact on Repeat-Purchase Loyalty Patterns, *International Journal of Marketing*, 14(5), 473–486; Mark D. Uncles, Grahame R. Dowling, and Kathy Hammond (2003), Customer Loyalty and Customer Loyalty Programs, *Journal of Consumer Marketing*, 20(4), 294–316; Jorna Leenheer and Tammo H.A. Bijmolt (2008), Which Retailers Adopt a Loyalty Program? An Empirical Study, *Journal of Retailing and Consumer Services*, 15(6), 429–442.
22. See Adrian Payne (1994), Relationship Marketing: Making the Customer Count, *Managing Service Quality*, 4(6), 29–31.
23. Adrian Payne (1994), Relationship Marketing: Making the Customer Count, *Managing Service Quality*, 4(6), 29–31.
24. Jay Kandampully and Liam Butler (2001), Service Guarantees: A Strategic Mechanism to Minimise Customers' Perceived Risk in Service Organizations, *Managing Service Quality*, 11(2), 112–120.
25. Ray Pelletier (2005), *It's All about Service: How to Lead Your People to Care for Your Customers*, Hoboken, NJ: John Wiley & Sons.

5

People, Technology, and Usability: An Ergonomic Perspective

In this chapter we exemplify and discuss a macro-perspective of people in nature and people and technology. We emphasize the need for service design to strive toward a model of harmony or balance between people-activity and the natural world. We feel this is a reasonable proposal. After all, nature itself is founded on a system in balance. Ancient so-called primitive societies lived, to a large extent, in balance with nature and its ecology. In our own understanding of the concept of ergonomics, we look simultaneously at ecology in nature and its inherent balance. The science of ergonomics focuses on human intervention often to the detriment of a natural balance. This is ergonomics from a macro-perspective. A more general understanding of ergonomics aims to create a balance between people as individuals or as groups on one side and on the other, technology from simple work tools to complex large technological systems. We take ergonomic theories and concepts as our springboard for service design.

Ergonomics: A Brief Overview

Ergonomics is a cross- and interdisciplinary science covering all different areas of human sciences, including some aspects of social science, as well as different areas of technology and technological applications.[1] Ergonomics can be subdivided into many different areas. We focus on four distinct areas: power ergonomics, informational ergonomics, environmental ergonomics, and organizational ergonomics. Power ergonomics deals with humans as an energy/power resource. From this point of view, ergonomics is related to the design of simple tools, for example, knives and forks, hammers, saws, screwdrivers, hand drills, and axes. In this category, we also include simple electromechanical hand tools, such as electrical hand drills, and household tools, such as vacuum cleaners and dish washers. Informational ergonomics is related to psychology, pedagogics, and sociology. But we emphasize that here there is only a relationship, and informational ergonomics is not a direct part of the overall discipline (as, for example, a university department of psychology). Practitioners of informational ergonomics (called ergonomists) need a good scientific understanding of humans as information processors, and different types of cognitive processes. They also need knowledge of the development and deterioration of these different types of processes in different types of work situations. In this context work can be both paid employment in industrial or office settings and related to human activities in the context of consumption (for example, at home) or leisure (for example, in outside venues, as when attending a concert). An information ergonomist needs to understand the interrelationship between human information processing and cognition in relation to information processing in technologically controlled processes. An example here would be a computer or different types of micro-computers and related embedded control systems.

Environmental ergonomics deals with the wide field of creating a kind of micro- or macro-climate around human beings to create comfort and physical balance in relation to the surrounding environment. In practice this relates to different types of clothing, air conditioning and ventilation systems, acoustical protections, and physical enclosures, such as buildings. Environmental ergonomics aims to create a balance in relation to the thermal environment and different forms of movement (for example, vibrations). Another feature relating to environmental ergonomics is the provision of the support systems for resting and working, for example, tables, chairs, and beds.

Organizational ergonomics is related to areas of information processing between groups of people (including the use of information processing devices such as mobile devices, desktop computers, and handheld computers) combined with multifunctional devices to function as a handheld computer terminal interlinking to centrally located computers. Included as part of this system are control rooms interacting with other control rooms in local, national, or international computer systems and all parts of the system capable of interacting with each other part. Sometimes these are a part of Internet systems or special purpose computer systems for long-distance information technology (IT) and information and communications technology (ICT) (for example, via satellite systems). These perspectives of informational ergonomics are developing at a rapid pace. This development is closely interlinked to an understanding of individual cognition and interaction with very large technological information processing components and very large supercomputers. From this perspective the context of organizational ergonomics is still in an early phase. Ripe for future development is advanced interaction on the Internet. This form of activity in a local society or on a global scale is an area of ergonomics. The human ability to handle, for example, large economic systems and the highs and lows of global economic systems is, to a large extent, dependent on this understanding of the human aspect of ergonomics and an understanding of technological systems of information handling (for example, as needed in processes of trading). Today, with interconnected macro-information systems, small errors in information handling can have enormous ramifications. On the other hand, information handling in a modern motor vehicle with many built-in electronic service systems (different types of embedded control systems) can also give rise to disturbances and accidents due to a lack of organizational ergonomics (i.e., a lack of understanding between the local service system and its coordination with the cognition of the human operator). These micro- and macro-ergonomic perspectives have many parallels to our understanding about overpopulation of our planet and the roles of human activity in building up macro-technological systems such as energy supply (e.g., nuclear power, wind power, water power) related to new types of energy-related consumption for heating, air conditioning, and transport (whether public or private). Seemingly, humankind currently lacks a good understanding of the ways in which human activities are affecting the balance in nature.

Fundamentals and Criteria of Ergonomics and Usability

Ergonomics focuses on the interrelation between humans and complex technological systems. The overall purpose of ergonomics is to contribute to a balance between people and technology. Figure 5.1 illustrates the relationship between people and technology in low and high levels of automation.

Although the sum of people and technology is the same in (a) and (b), levels of production output are likely to be higher in (b) to reflect the higher employment of technology in the work process. However, production in (a) may be related to a specialized human skill or craft, such as handcrafted embroidery, where the input of human expertise is high and limited use is made of technology to support the human labor. Figure 5.2 illustrates the relationship between people and technology where automation is at a very high level and where automation is nonexistent.

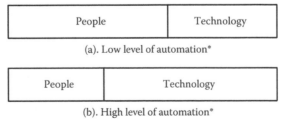

(a). Low level of automation*

(b). High level of automation*

* In each instance the sum of people + technology is the same.

FIGURE 5.1

People and technology in automation. (From Toni Ivergård and Brian Hunt (2010), New Approaches in the Design of Future Control Rooms, invited plenary paper for the International Control Room Design Conference (ICOCO), Paris, October 25–26.)

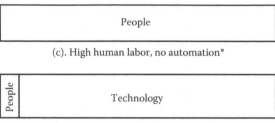

(c). High human labor, no automation*

(d). High automation, low human labor*

* In each instance the sum of people + technology is the same.

FIGURE 5.2
People with and without technology. (From Toni Ivergård and Brian Hunt (2010), New Approaches in the Design of Future Control Rooms, invited plenary paper for the International Control Room Design Conference (ICOCO), Paris, October 25–26.)

The work environment represented in (c) has highly intensive use of human labor and no technology to support this labor. Nowadays, there are very few examples of work that use no technology whatsoever. Even a simple farm implement is a form of technology. The work environment represented in (d) employs technology to a large extent and has very low human labor input, although the human input is likely to be highly skilled to manage and maintain the high levels of technology in the work process. Modern production lines resemble this format. On a guided tour of a vehicle manufacturing plant the visitor sees many robots and robotic-guided processes (such as transporting parts to the production line). The robots (nicknamed steel-collar workers) are employed in work traditionally done by technicians and crafts-people, such as welding, riveting, and installing components to construct the finished vehicle.

Ergonomics: Aiming for Balance

Ergonomics deals with micro-perspectives from one person and one work tool up to complex organizations, whether these are national or international. The key aims of ergonomic investigation in this area are to create balance and avoid unwanted disturbances, and in the worst case, accidents or catastrophes. The ergonomic component and its approach provide a key to success in realizing this process of balance. Balance is, for example, critical in helping humankind solve the problem of climate change and global warming. The lack of a systematic approach, as suggested by the use of ergonomics, frequently results in suboptimizations. The problems of global warming or climate change cannot alone be solved by one individual science, such as, for example, meteorology. The problem of global warming can only be handled with sound, concerted, interdisciplinary efforts. Nor will ergonomics per se fully resolve issues of global warming. But ergonomics is an example of an interdisciplinary approach that can handle human technology problems that require an interdisciplinary approach.

The reason for the current lack of success in the application of ergonomics in the macro-area is attributable to the fact that traditional research institutions and universities are in essence disciplinary in an old-fashioned way. The old division of science into different faculties is completely outmoded. Understanding complex problems can never be understood in a good way within a single science. Professor Albert Einstein reminded us that problems cannot be solved with the same thinking that created them. Invariably, whether it is the detail or a macro-perspective view of a problem, better results can be achieved by using broad interdisciplinary methods and involving people from diverse disciplines and perspectives. (We belabor this point as we feel that there are lessons here for the management of service.)

On a micro-level ergonomics has proven to be very useful. When we look at problems of power ergonomics or information ergonomics, we use different types of criteria to evaluate and support the design work. Examples of criteria used include factors that affect quality of life at work and leisure. A typical list of criteria is:

1. Health, safety, and security
2. Functional, durable, easy to use
3. Comfortable versus cozy
4. Aesthetic, i.e., has a good appearance from a subjective/objective/ normative perspective
5. Simplistic versus complex
6. Pleasurable and stimulating (including sexuality)
7. Passion, desire, and craving
8. Motivation (i.e., a driving force for behavior)

Sometimes the word *usability* is used in a meaning similar to *ergonomics*. Most commonly *usability* is used in the meaning of ease of use of ICT products. A more formal definition can be found in the work of Brian Shackel, which encompasses the goals of effectiveness, learnability, flexibility, and attitude.[2] These attributes are encompassed in a short question: Will the users be able to work the technology successfully? There are also a number of techniques that have been developed in the field of ergonomics. The physical bodily dimensions of people presented in computer databases or in the former human scale is one example of the design of workplaces. Another important tool for ergonomic design work is the allocation of functions. Figure 5.3 shows the use of allocation of functions.

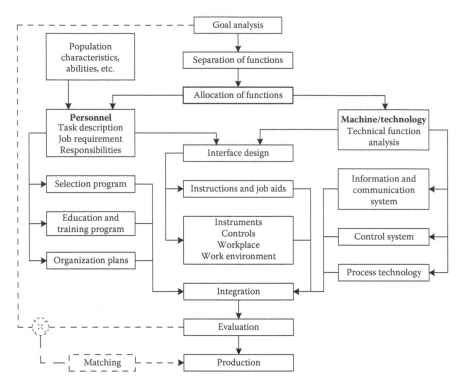

FIGURE 5.3
Allocation of functions. (From Toni Ivergård (1981), *Information Ergonomics*, Lund, Sweden: Studentlitteratur.)

Allocation of functions can vary from different users and designers. The methodology was created during the 1960s by ergonomists in the UK and the United States. Singleton (1974) made an elegant description of the use of allocation of functions.[3] In his unpublished doctoral thesis Toni Ivergård (1972) described retailing from the development perspective of allocation of functions and predicted the move from manually orientated retailing (shop assistants serving behind counters) to highly automated retailing (bar code scanning and automated readers of customers' purchases). In his predictions of the roles and purposes of technology, he came very close to what today is referred to as Internet shopping.

Motives for Physical Design as a Part of Excellence in Design and Practice

Ergonomics as a science seeks to adapt technology to fit people. This fundamental theme pervades our thinking throughout this book. In this

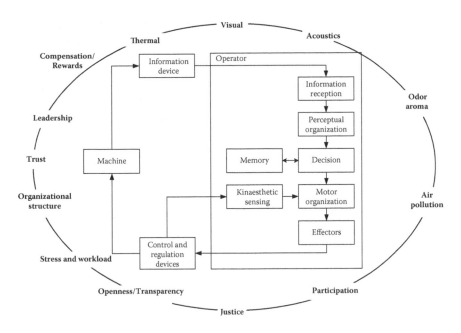

FIGURE 5.4

Environment: physical and psychological/social. (From Toni Ivergård (1981), *Information Ergonomics*, Lund, Sweden: Studentlitteratur.)

current chapter we place excellence of service in a context of macro- and micro-perspectives. This helps in comparing these two perspectives and aids analysis of their strengths and weaknesses. In other sections of this book we discuss in more detail how to obtain a good ergonomic design in order to achieve a high level of service excellence. By definition, good ergonomic design fits all different types of people. Furthermore, ergonomics deals with all different aspects of people-related perspectives. For example, ergonomics considers the characteristics of people, such as their physiological dimensions, their physical strength, and also their psychological and mental perspectives. In terms of context, environmental factors such as lighting, acoustics, and air pollution are also of relevance. Another critical factor to consider is workload. Other important issues relate to the quality of work and levels of experience of employees. Figure 5.4 shows an example of people-technology interaction in relation to information ergonomics with the addition of a number of environmental factors to illustrate the inclusion of environmental ergonomics.

In using ergonomics in the design for people at work, at leisure, or as consumers, a number of criteria are relevant. To ensure good design, these criteria should be adhered to and fulfilled either wholly or to varying degrees. Criteria vary according to context and situation, and can also vary depending on the objective. The following criteria need to be highlighted: health and safety, well-being and comfort, durability and sustainability, and pleasure and happiness.⁴ Sometimes, and particularly in service for leisure and pleasure pursuits, other factors of relevance might be passion, lust, and addiction.⁵

Safety and health are paramount in a number of industry sectors. Transportations systems (airlines, railways, bus services) make a point of emphasizing passenger safety. Airlines are legally obligated to demonstrate safety features of the aircraft before takeoff. Cabin crew announce that all passengers (including frequent fliers) should pay attention to their safety demonstration as safety features (such as the location of emergency exits) vary on different aircraft. Frequent fliers consider certain seats in the aircraft to be safer than others and when booking and at check-in, request their preferred seat (aisle seat, near the exit, over the wing, near the front). Air travel is acknowledged statistically to be the safest form of travel. It is said that airline passengers are most at risk in the taxi ride to the airport. The hospitality industry also emphasizes health and safety of customers as a duty of care. Arguably, in some establishments health and safety of employees may be secondary to other concerns, and in others practically nonexistent. We've seen some kitchens and back stairs (customer-free) areas on several continents where work environments seem positively dangerous. It is said that the kitchen is the location of 90 percent of accidents in the home. In industrial and professional kitchens the accident risk is greater, and there tend to be more people present and involved in a greater number of work tasks simultaneously and under time pressure. Management tends to set strict rules and routines in an effort to prevent accidents in these workplaces.

Somewhat related to health and safety are well-being and comfort. Unlike in transportation systems, failures in well-being and comfort are unlikely to be life threatening, but nonetheless, breaches in levels of well-being and comfort can raise the emotions of users and, in some cases, lead to accidents. And, surprising as it may seem, issues of health and safety and well-being bring public relations crises of almost equal importance. Health and safety influences the operations of many industries, including those that serve food (hotels, restaurants, cafés, clubs, universities and schools, public transportation services). When food causes illness or worse, various governmental agencies and public bodies mobilize resources to address public concerns. For the companies involved, declining public confidence forces them to take steps to manage the crisis.⁶ In a range of industries and settings a critical success factor (CSF) is safety of the customer.

Ergonomics as a philosophy is related to humankind's understanding of ecology and the harmony found in the cyclical processes of nature. Nature's processes are in balance. Human production and consumption intervene in this natural balance. The enormous and relatively recent population increase on our planet strongly threatens nature's balance and harmony. From a limited perspective, ergonomics attempts to learn from the balance and harmony that are found in nature. When considering ergonomics of systems for individuals, it is conventional to refer to micro- or macro-ergonomics. Macro-ergonomics relates to groups of people in relation to a larger system/organization. A macro-perspective of ergonomics deals with a group of people interrelated with some kind of complex technological system. Ergonomics in a micro-perspective deals with one or a few individuals interacting with a specific part of a technology.

Usability/Ergonomics Related to Different User Groups

The ultimate *aim* of any type of ergonomics is to include all different types of user groups. The ultimate *vision* is to make designs to fit all. If possible, we should include all human beings irrespective of age, physical abilities or disabilities, races or nationalities, gender and lifestyle preferences. An important starting point is to use basic anthropometrics (physical measurements of the human body). Other types of data of humans are the function and capabilities of our sense organs and strength and mobility of our muscles. Obviously, consideration should also be given to features of human cognition and learning abilities and restrictions. The ergonomic designer will on this basis define a need to use technology to supplement human restrictions. This is a role of the allocation of functions. People with low or restricted personal mobility need a safe and functional design of streets, stairways, and other access routes in order to facilitate accessibility irrespective of the individual's difficulty in moving around. For example, a city and a building should be designed to facilitate the use of wheelchairs and have handrails to aid access for people who have physical difficulties in personal mobility. People with visual impairments should be facilitated in the environment with suitable aids. These aids could be in the form of tactile sensors or modern equipment, for example, in the form of acoustic or electronic signals or built-in radar equipment that the individual can perceive through the use of other senses, such as touch or hearing. Hearing aids have over recent years been greatly developed to supplement impaired hearing. These are just a few examples relating to allocation of functions and how technology combined with innovations can supplement people at leisure and work.

Rapidly aging societies need a completely new approach. Many parts of the world have seen a very rapid increase in the numbers of aging people, while at the same time there are apparent reductions in the numbers of people who comprise the productive age group (the workforce). This is an incongruous situation that is potentially unsustainable. Over time this situation can lead to economic stagnation and undue social and financial burdens on society. The productive proportion of people between the ages of twenty-five and fifty-five cannot be expected to teach and train the group below the age of twenty-five while at the same time shouldering the responsibility for caring for the ill, infirm, diseased, and elderly who are unable to manage on their own. In this scenario very few people will remain to provide the means of production for the society. If steps are not taken to rectify this, it will be the actual situation in a number of countries within the next half century.[7] A new approach to designing our living environment can make it possible to a much larger extent for the elderly and the infirm to take care of themselves.

Self-care will also provide and give provisions for a healthier type of aging. This is the basic concept behind prevention of illness and disease and is superior to the concept of curing. Of course, this is not to say that sometimes curing is valuable and necessary. But many times the progression from illness to cure becomes a vicious circle. In the near future people above 100 years old will comprise quite a large group. However, it would be naïve to expect too much productivity from this 100+ age group, especially as they most likely feel that they have contributed to their societies earlier in their lives.

Information and communications technologies (ICTs) will be a useful tool to facilitate usability for inclusion of more or less all people in a rather natural way of living. But sadly enough, new technologies in the form of mobile devices seem mainly to focus on cognition and learning of the young. Compared with older generations (people aged twenty-five years and above), young people have an extremely good short-term memory and related learning abilities and seem able to retain these attributes even in stressful and noisy conditions. The elderly (those over the age of forty-five) are very strong on long-term memories and can make references of lifelong experiences stored deep in the brain. But when attempting to learn new things, elderly people tend to need more time for processing and adapting, part of which is spent on time-consuming trial-and-error processes. In general, older people have slower reaction times and lower levels of dexterity. Their use of keyboards and the positions and speed of finger movements are less precise. Posture and balance deteriorates with age. Elderly people with an active physical life can often compensate and reeducate their physical impairments and deteriorations, in essence retraining muscle memory. Impaired vision and hearing is a natural part of aging. As long as these impairments are not too severe, these two impairments can be overcome with devices such as spectacles and hearing aids. The new world of ICT must include the aging population and their cognition. In this way, the elderly can remain fit and physically active up to much higher ages. Even mental abilities can be developed. The elderly can then continue to contribute to the production of their society. They can also increase their quality of life such that life becomes more pleasurable and fun for society as a whole.

In the time of the Vikings (approximately from the eighth to the eleventh centuries) there was a practice of people committing suicide when they became too old to fulfill active roles in their society. Today, we are more sophisticated and should have other values and possibilities for the process of aging. But to a large extent this depends on what we are doing in society of today in terms of urban planning, building design, and forward planning to provide for an active productive life of the elderly. As was concisely stated in a United Nations report on aging:

> The challenge for the future is "to ensure that people everywhere can grow old with security and dignity and that they can continue to participate in social life as citizens with full rights." At the same time "the rights of old people should not be incompatible with those of other groups, and reciprocal intergenerational relations should be encouraged."[8]

The World Is Getting Older

Toward the end of the twentieth century, the population of our planet was reported to have passed 6 billion, and by 2005 it had reached 6.5 billion.[9] In its 2005 report, the Department of Economic and Social Affairs of the United Nations Secretariat reported that global population grows at an annual rate of 1.2 percent. A twelve-year period (1987–1999) saw global population grow by another 1 billion.[10] This twelve-year period is the shortest time taken to add another 1 billion people to the population on our planet. An earlier report prepared by the Population Division of the United Nations indicated four critical indicators.[11] These are population aging is unprecedented and is without parallel in human history; population aging is pervasive—this phenomenon is global and every person on the planet will be affected; population aging is enduring—this is the future and humankind will not revert to young populations; and population aging has profound implications for many facets of human life.

Several recent official reports remind us that populations are aging fast in many countries around the world. A report by the Organization for Economic Cooperation and Development (OECD) states that by 2050 some 2 billion people will be aged over sixty years. These data have implications throughout these aging societies. The phenomenon of increasing aging populations needs decisions to be made at government levels concerning health-care and social support and related resources, such as budgets, care facilities (specialized equipment and appropriately trained care professionals), and frameworks for social and environmental infrastructures. Some of these decisions also need to be made at regional levels, for example, acquiring land and space for the design and construction of housing and accommodation and public transport networks. At present this phenomenon is confined mainly to developed nations such as Germany, Italy, Japan, and the United States. However, countries in the developing world are not much more than one generation behind. The elderly population (generally defined as people aged sixty-five or older) occupy increasingly higher percentages of total population. In certain developed countries this percentage of the total population is predicted to acclerate.

The population of developed countries is aging fast, and the developing world is lagging behind at most two or three decades. In OECD member countries the fastest-growing segment of the population is older adults (those aged sixty-five and over).[12] Aging represents one of the most important challenges for the OECD member countries.[13] It is projected that by 2030 around 25 percent (i.e., one in every four persons) of these countries will be aged sixty-five years or older.[14] Currently, less than 60 percent of the populations in OECD countries is in paid employment. Employed workers in the age range from twenty-five to forty-nine years of age account for around 75 percent of the OECD population. This is predicted to change rapidly over the coming decades so that by 2050 the ratio of employed to unemployed could be 1:1.[15] This gives rise to a paradox: as the proportion of elderly people in the OECD countries *increases*, the share of the working population is predicted to *decrease*.[16] Governments will therefore need to tackle some major issues and face difficult decisions, not the least of which relates to national income. Fewer persons employed in the workforce will contribute their labor production to gross domestic product (GDP). As per capita work output productivity and income tax contributions fall, there will be fewer people providing national revenue to support nonproducing members of a population (such as the infirm, the sick, and the disabled). An aging population thus represents a threefold risk to national economies at the fiscal, social infrastructure and at the level that affects the well-being of individual members of the population. The ramifications will be both political and social.

Peter Part, chairman of the Working Group on Aging Populations and Sustainability of the Economic Policy Committee (EPC) of the European Union, states:

> The ageing of the population is becoming a growing challenge to the sustainability of public finances in the EU Member States. The increase of the ratio between the number of retirees and the number of workers will amplify expenditure on public pensions and health and long-term care and thus puts a burden on maintaining a sound balance between future public expenditure and tax revenues.

Not only is a rapidly aging population predicted to put into question the possibility of sustained national economic growth, but it also threatens the financial health of the aged themselves.[17] A report by the EU captures the gravity of an aging population on economies: "An ageing population raises challenges for our societies and economies, culturally, organisationally and from an economic point of view."[18] The projected population of the EU in 2060 will be 517 million citizens (an increase of approximately 15 million from the current population). By 2060 it is predicted that about one-third of these citizens will then be aged 65 or over.

The population of Japan is aging faster (at an extraordinary rate) than that of any other OECD member country.[19] The percentage of the population over the age of sixty-five is predicted to almost double in size from around 25 percent today (2014) to around 45 percent by the year 2050.[20] Japanese law stipulates retirement at sometime between the ages of sixty and sixty-five, and because of the post-WWII baby boom, in the next few years Japan will see a large number of the elderly leaving its workforce.[21] The large number of retirees will put pressure on the Japanese healthcare industry, especially the training of healthcare professionals in sufficient numbers to cope with the large population of aged people.[22]

As various official reports affirm, different nations are at different stages of aging. Countries in the developed world (such as OECD member states) have for some time seen the onset of their aging societies and the increasingly large numbers of their citizens aged sixty-five or over. Developing countries are just beginning to see changes in their age demographic. As countries are aging at a different pace, so will the timeframe at which governments and societies prepare and adapt. For all countries, changes in their aging demographic will arrive sooner or later. Nature's biological clock does not run backward. Societies are not predicted to regress to infancy or adolescence. Countries (such as those in the developing world) whose full trajectory of aging has yet to appear will have shorter timeframes in which to design government policies and refine their social infrastructures.

Aging Populations and Sustainable Work Life

In the following section we discuss two alternative hypotheses for an active and high-quality contribution by a workforce of the age range from sixty to eighty years old. We primarily focus on the effects of ICT in relation to an aging workforce. In this area there appear to be two key issues. In essence, do new and emerging technologies (specifically ICT) *support* an aging population in an active participation in the workforce and life overall, or are these technologies setting *obstacles* for active participation by an aging population?

Over the past several decades there has been a research and development focus on ICT, and in particular its relation to issues in human factors (HFs)/ergonomics, with an emphasis on "information ergonomics."[23] The past decade has seen more work conducted on developing the concept of usability as a subsection of HFs/ergonomics to be more specific on people-ICT interactions. There has also been a very large quantity of research and literature directly or indirectly related to workplace control rooms (CRs) and command and control centers (CCCs).[24] This research can to some extent help us to understand different kinds of time-related perspectives of human employees' interactions with and usage (over time) of different kinds of ICT. Mobile "smart" ICT as a development from mobile phones has partly had a similar process of adding artificial intelligence (AI). But the process is much more dissipated, and it would not be unfair to use the word *disparate* to describe some areas' development of smart handheld ICTs.

It is somewhat easier to track down and explain the process of development over a number of decades in the area of CR/CCC development. But this is not to suggest that this is perfect, as initial perceptions and understandings may be misleading. At first, the usage of smart handheld ICTs might be counterproductive for an aging workforce. Further research is needed in the key areas of human cognition, psychology, and neurophysiology, both in a very broad understanding and in their relation to different generations of ICT and future new possibilities for an elderly workforce. Over the past few decades and on several continents ICT has become an accepted part of daily life for vast numbers of people, particularly for the young. The spread of ICT is currently growing very fast in Asia and in many parts of Africa.[25] Arguably, a key issue related to this application is: Will the next generation of ICT be an obstacle, or will it be an opportunity for a workforce aiming to work at a higher age (60 to 70+)? Here an opportunity would include technology for work purposes (for example, technology to enable people to work more effectively). The so-called instrumental activities of daily living (IADLs) can be improved by design changes to alleviate limitations in abilities to perform some (mainly physical) activities.[26] The European Commission (EU) has carried out and initiated a large number of projects about ICT and aging.[27]

In their book *Technogenarians* Kelly Joyce and Meika Loe describe an elderly person's experiences of health, illness, science, and technology. This book presents an impressive theoretical and empirical understanding of the biomedical aspects of aging bodies, minds, and emotions, and the rise of gerontechnology industries and professions. The book combines two scholarly areas: science and technology studies and sociology of aging and health, and illness. The authors and their contributors investigate the elderly user of technology, and their findings are indirectly related and relevant to the elderly at work.[28] A similar neologism, *gerontographics*, was coined by Professor George Moschis to describe a scientific approach to analyzing and targeting the mature market.[29]

Many new technologies are already on the way to the market: for example, haptic technologies (in the hands and in the body as a whole) and ones that use auditory and odorous inputs. Large multitouch display screens (which may also be in 3D) are useful in environments for senior management and in executive boardrooms. These screens are also suitable for the display of information for education and public venues. Processes such as data warehousing and cloud technology are also interesting and relevant developments. Writer and TV presenter David Pogue describes the new superfast Windows on an iPad as an example of the "long promised world of 'thin client' computing."[30] But, this is all very new, and in general, there is a lack of experiences (such as personal case histories) and research-based data related to the new ICTs, particularly in relation to aging users.

An Aging Workforce and Technology

It is a fact of life that normal aging processes affect human abilities and functions. There is a marked deterioration in physical attributes, for example, strength, speed, precision, and tactile (manual) dexterity. Also, there are reductions in one's mental abilities, such as different forms of cognition and short-term memory and sensory abilities such as vision, hearing, and touch. In addition, aging tends also to lead to reduced health in general. To a certain extent this trend is being reversed.[31] In terms of technology use, a key issue is: Will an average sixty-five-year-old person be supported by the new generation of ICT (e.g., multitouch and haptic facilities), or will technological development be an additional impediment in the natural process of aging? Obviously, there are many potential obstacles, some of which are evident even now. There are likely to be a number of conflicts and contested issues. Reduced short-term memory might be compensated by motoric memories/abilities. Given the speed with which some societies are aging, there is an urgent need for cross-disciplinary research into the changes of human skill and abilities experienced by aging and the technological development of the new generations of ICT. Most likely there will be a need for personalized solutions and supplementary aids.

Managing and Assuring Ergonomics and Usability

The approach of a modern understanding of ergonomics can bring enormous value for the development of our society. As we have described in this current chapter, ergonomics defines a new approach for how technology can contribute to human life and the survival of the natural world. Science and research need to be based on interdisciplinary understandings and work methods. The old division and segmentation of research and science is outdated. In ergonomics we often talk about systems ergonomics and a systems approach to ergonomics. One of the current authors (TI) applied an ergonomic approach to labor economics during his time as a labor director of the Swedish Labor Administration. This ergonomic approach successfully defined new ways of achieving a good match between supply and demand of work competencies and skills.[32]

In systems ergonomics, there has developed a very good understanding of information to prevent accident and catastrophes. For further development, there needs to be many more audits and other types of follow-up in ergonomics activities. Recently we have been involved in the evaluation control rooms in a very large petrochemical industry complex in Thailand. We note the rapid increases in the design of control rooms and control centers. But a continued, and major, problem seems to be the reluctance to engage in incorporating in the design relevant processes and systems of evaluation and follow-ups. This reluctance inhibits progress. Many large organizations are today using ergonomics in the design of their control rooms for business processes and logistics. However, ergonomics is only being used in a very limited sense in relation to display, design, furniture design, and such. A systems approach tends not to be included.

In the service industry, we can see a trend of making increased use of ICT in its broadest sense. But currently, the usage is at a rather primitive stage, even if there has been an explosion in the number of applications. While customers use mobile technologies, the industry seems to focus on using available technologies (e.g., the Internet) for marketing, sales, and booking. This is widely used for hotels and accommodations, and in ticketing for airlines, buses, and railways. For some time customers have used mobile devices to make reservations, booking, and payment. However, the technology is still very limited in its use for the purposes of organizational integration and usage within the real core business activities. There are noteworthy exceptions. At Gyros, a Greek restaurant recently opened in Doha, customers order from menus displayed on iPad computers. Waiting staff take customers' order on their iPhones. Orders placed by the customers go straight to the kitchen team, thus speeding up the process of getting prepared food to the tables. Aloft hotels in New York and California offer keyless check-in facilities that allow guests to access their rooms without being processed at the reception desk. After a trial period of three months, the technology is planned for all Aloft and W hotels, which are part of the Starwood Hotel and Resort group.[33]

New technologies may present problems because many of the potential users, such as the elderly (the 35+ age group), will have difficulties in using these multipurpose systems designed mainly to meet the needs of a younger generation of users. But at the same time, the elderly are one of the main target groups for this form of usage, and financially, they are often more than able to pay for this type of service if it provides a high level of usability. Currently middle and top managers in organizations tend to rely on their younger assistants and secretaries to interface with technologies on their behalf. Better-designed technology (from a usability point of view) will have the potential to provide better service and increased sales revenues for the service provider (hotels, airlines, intermediaries, ticketing and travel agents, etc.).

The oncoming new facilities in this area will have the potential of a much higher level of usability also suited to elderly users and the very old.

Mobile communication tools can be very useful service equipment for the very old (the 70+ age group), especially those who are physically immobile. Mobile devices can improve comfort and daily security and also provide safety communications in case of accidents. A future extensive use of these handheld mobile devices will require some kind of standardization, and also in-built facilities for living in organizations in hospitals and in general in the town planning. To achieve this in a good way there is a need for a new type of technical-human ergonomical planning in relation to building infrastructure design.

Endnotes

1. For detailed discussions of issues in ergonomics, see, for example, Stephen J. Guastello (2013), *Human Factors Engineering and Ergonomics: A Systems Approach*, Boca Raton, FL: CRC Press; Gavriel Salvendy (ed.) (2013), *Handbook of Human Factors and Ergonomics*, Hoboken, NJ: John Wiley & Sons; Tonya L. Smith-Jackson, Marc L. Resnick, and Kayenda T. Johnson (2013), *Cultural Ergonomics: Theory, Methods, and Applications*, Boca Raton, FL: CRC Press; R.S. Bridger (2008), *Introduction to Ergonomics* (3rd ed.)., Boca Ratan, FL: CRC Press.
2. Brian Shackel (1991), Usability—Context, Framework, Definition, Design and Evaluation, in Brian Shackel and Simon Richardson (eds.), *Human Factors for Informatics: Usability*, Cambridge: Cambridge University Press, Chapter 2; Brian Shackel (1997), Human-Computer Interaction: Whence and Whither? *Journal of the American Society for Information Science*, 48(11), 970–100; Brian Shackel (2009), Designing for People in the Age of Information, *Interacting with Computers*, 21(5–6), 325–330.
3. W.T. Singleton (1974), *Man-Machine Systems*, Harmondsworth, UK: Penguin Books.
4. See discussions in Peter Hasle and Per Langaa Jensen (2012), Ergonomics and Sustainability: Challenges from Global Supply Chains, *Work*, 41, 3906–3913; Giuseppe Di Bucchianico, Antonio Marano, and Emilio Rossi (2012), Toward a Transdisciplinary Approach of Ergonomic Design for Sustainability, *Work*, 41, 3874–3877; Justine Arnoud and Pierre Falzon (2012), Shared Services Centers and Work Sustainability: Which Contributions from Ergonomics? *Work*, 41, 3914–3919.
5. See, for example, Waldo C. Klein and Carol Jess (2002), One Last Pleasure? Alcohol Use among Elderly People in Nursing Homes, *Health and Social Work*, 27(3), 193–204.
6. See Michael Regester and Judy Larkin (2008), Risk Issues and Crisis Management in Public Relations (4th ed.), London: Kogan Page; Susan Miles, Mary Brennan, Sharron Kuznesof, Mitchell Ness, Christopher Ritson, and Lynn J. Frewer (2004), Public Worry about Specific Food Safety Issues, *British Food Journal*, 106(1), 9–22; Regina E. Lundgren and Andrea H. McMakin (2013), *Risk Communication: A Handbook for Communicating Environmental, Safety and Health Risks* (5th ed.), Hoboken, NJ: John Wiley & Sons.

7. See Roald Lee, Andrew Mason, and Daniel Cotlear (2010), *Some Economic Consequences of Global Aging*, Washington, DC: World Bank; Daniel Cotlear (ed.) (2011), *Population Aging: Is Latin America Ready?* Washington, DC: World Bank.

8. United Nations (2001), *International Strategy for Action on Ageing 2002, draft text proposed by the chairman of the Commission for Social Development acting as a preparatory committee for the Second World Assembly on Ageing at its resumed first session, New York, December 10–14.*

9. United Nations, Department of Economic and Social Affairs, Population Division, *The World at Six Billion*, New York: United Nations, 1999.

10. United Nations (2005), *Population Challenges and Development Goals*, New York: United Nations, p. 5.

11. United Nations, Department of Economic and Social Affairs, Population Division (2005), *World Population Aging: 1950–2050*, New York: United Nations.

12. Thirty-four countries are members of the OECD. In December 1960 there were twenty founding members. Since then fourteen more countries have become members. OECD member states are Australia, Austria, Belgium, Canada, Czech Republic, Denmark, Estonia, Finland, France, Germany, Greece, Hungary, Iceland, Ireland, Israel, Italy, Japan, South Korea, Luxembourg, Mexico, Netherlands, New Zealand, Norway, Poland, Portugal, Slovak Republic, Spain, Sweden, Switzerland, Turkey, United Kingdom, and United States.

13. OECD (2006), *Live Longer, Work Longer*, Paris: OECD Publications, p. 9.

14. OECD (2001), *Ageing and Transport: Mobility Needs and Safety Issues*, Paris: OECD Publications.

15. OECD (2006), *Live Longer, Work Longer*, Paris: OECD Publications, pp. 9–10.

16. See Milan Vodopivec and Promoz Dolenc (2008), *Live Longer, Work Longer: Making It Happen in the Labor Market*, Washington, DC: World Bank.

17. Milan Vodopivec and Promoz Dolenc (2008), *Live Longer, Work Longer: Making It Happen in the Labor Market*, Washington, DC: World Bank.

18. *The 2012 Aging Report: Economic and Budgetary Projections for the 27 EU Member States (2010–2060)*, joint report prepared by the European Commission (DG ECFIN) and the Economic Policy Committee (AWG), Brussels: European Commission.

19. Eiiichi Oki and Hisakazu Matsushige (2011), Japan's Aging Population and Its Silver Care Industry, *Seri Quarterly*, October, 35–45.

20. Jun Inoue (2013), Healthcare: The Case of Japan, *Migration Letters*, 10(2), 191–209.

21. Eiiichi Oki and Hisakazu Matsushige (2011), Japan's Aging Population and Its Silver Care Industry, *Seri Quarterly*, October, 35–45.

22. Jun Inoue (2013), Healthcare: The Case of Japan, *Migration Letters*, 10(2), 192.

23. Toni Ivergård (1982 [1969]), *Information Ergonomics*, Lund/London: Studentlitteratur/Chartwell-Bratt. (This book was published first in Swedish (1969) and later in English (1982).)

24. See Toni Ivergård (1969), *Informationsergonomi, Rabén & Sjögren*, Stockholm (in Swedish); Toni Ivergård (1982), *Information Ergonomics*, Lund/London: Studentlitteratur/Chartwell-Bratt; Brian Shackel (1981), Man-Computer Interaction, *Ergonomics*, 12, 485–500; Martin G. Helander, Thomas K. Landauer, and Prasad V. Prabhu (eds.) (1997), *Handbook of Human-Computer Interaction*, Amsterdam: North-Holland Publishers; John R. Wilson and Nigel Corlett (eds.) (1998), *Evaluation of Human Work*, London: Taylor & Francis; Toni Ivergård and Brian Hunt (2008), *Handbook of Control Room Design and Ergonomics: A Perspective for the Future* (2nd ed.), Boca Raton, FL: CRC Press/Taylor & Francis.

25. See Gudrun Wicander (2011), Mobile Supported e-Government Systems, PhD thesis, Faculty of Economic Sciences, Communication and IT Information Systems, University of Karlstad, Sweden. Also see Frank B. Tipton (2002), Bridging the Digital Divide in Southeast Asia, *ASEAN Economic Bulletin*, 19(1), 81–99; Poh-Kam Wong (2002), ICT Production and Diffusion in Asia, *Information Economics and Policy*, 14(2), 167–179; Anne Milek, Christophe Stork, and Alison Gillwald (2011), Engendering Communication: A Perspective on ICT Access and Usage in Africa, *Info*, 13(3), 125–141; Kallol Bagchi, Godwin Udo, and Peeter Kirs (2007), Global Diffusion of the Internet XII: The Internet Growth in Africa: Some Empirical Results, *Communications of the Association for Information Systems*, 19, 325–351.

26. See, for example, David Seidel, Kathryn Richardson, Nathan Crilly, Fiona E. Matthews, P. John Clarkson, and Carol Brayne (2010), Design for Independent Living: Activity Demands and Capabilities of Older People, *Ageing and Society*, 30(7), 1239–1255.

27. See, for example, Jan Snel and R. Cremer (eds.) (1994), *Work and Aging: A European Perspective*, London: Taylor & Francis.

28. Kelly Joyce and Meika Loe (eds.) (2010), *Technogenarians: Studying Health and Illness through an Aging, Science and Technology Lens*, Chichester, UK: John Wiley & Sons. Also see Toshio Obi, Diana Ishmatova, and Naoki Iwasaki (2013), Promoting ICT Innovations for the Ageing Population in Japan, *International Journal of Medical Informatics*, 82(4), 47–62.

29. George P. Moschis (1993), Gerontographics: A Scientific Approach to Analyzing and Targeting the Mature Market, *Journal of Consumer Marketing*, 10(3), 43–53.

30. David Pogue (2012), Windows Superfast on Your iPad, *International Herald Tribune*, February 23.

31. Stephen Katz (2001–2002), Growing Older without Aging? Positive Aging, Anti-Ageism, and Anti-Aging, *Generations*, 25(4), 27–32; Milan Vodopivec and Promoz Dolenc (2008), *Live Longer, Work Longer: Making It Happen in the Labor Market*, Washington, DC: World Bank; Paul Higgs, Miranda Leontowitsch, Fiona Stevenson, and Ian Rees Jones (2009), Not Just Old and Sick: The 'Will to Health' in Later Life, *Ageing and Society*, 29, 687–707.

32. See Toni Ivergård (2000), An Ergonomics Approach for Work in the Next Millennium in an IT World, *Behaviour and Information Technology*, 19(2), 139–148.

33. Hotel Chain Begins Offering Check-In by Smartphone (2014), *International New York Times*, February 5, p. 18.

6

Leading Organizations and Employees toward Service Excellence

Identifying Leadership

Identifying leadership has been likened to the story of the blind men and the elephant.[1] Originally from India, the story relates each of several blind men touching a different body part of the elephant (the trunk, the body, a leg, an ear). Being blind, none is able to see the animal in its entirety. The individuals form a mental image of the whole beast from one of the parts. The person who touched the elephant's trunk suggests that the animal is like a snake. Others suggest that the animal is like a wall (from touching the body), a tree (from touching a leg), and a fan (from touching an ear). Similarly, leadership then may be considered from any number of angles and, as a result, offer different perspectives. At the core of definitions of leadership is a sentence or two that describes the ability of one person to influence the behaviors of another person or group.[2] The relationship of the leader (the influencer) to the follower(s) (the influenced) is a necessary condition of leadership. Without followers, there is no leader.[3] As President Franklin D. Roosevelt (1882–1945) allegedly said, "It's a terrible thing to look over your shoulder when you're trying to lead and find no-one there." The ability to attract and then retain followers is a key attribute of a successful leader.[4]

According to some commentators, good leaders excel in four key areas: setting and affirming values and purpose, setting a vision and related strategies to achieve it, building coalitions of employees to execute strategies, and initiating and managing organizational change.[5] An organization's values and purpose serves as a blueprint for thought and actions. The vision set by an organization's leaders needs to be meaningful to insiders (managers, employees, and the executives themselves) as well as external stakeholders such as customers, suppliers, and investors. For the insiders a key aim is to attain unity of purpose—everyone's effort pushing in the same direction. For external audiences the vision helps demonstrate a unified organization and one that has a strong healthy culture. An organization with a strong healthy culture means that its members align their efforts toward a common goal. Workplace effort in alignment is one of three hallmarks of a positive corporate performance.[6]

It is a truism that an organization's culture defines its work practices. Cultures also exercise an impact (whether positive or negative) on the firm's performance over the longer term, which may include the organization's propensity to thrive, to survive turbulent environments, or to decay and cease operations. The more robust is an organization's culture, the more the organization can be predicted to survive when its business environments become turbulent and are subjected to rapid change. An organization with a strong healthy culture has harmony in its workforce in that there is a common consensus of effort.[7] Such organizations tend to have a highly motivated workforce. Organizations with a corporate culture where employees lack cohesion may be described as having a weak or unhealthy culture. In such cultural environments work effort is diffuse and tends to be disjointed. Organizations in this situation struggle to survive.[8] For example, the organization may have difficulties retaining competent employees or attracting and recruiting new competent employees.

A Leader's Key Tasks: Develop the Organization and Its People

Arguably, organizational leaders have four critical tasks: set a vision for actions and communicate this vision to relevant parties, communicate expectations for workplace tasks and behaviors, facilitate employees to grow both personally and professionally, and emphasize commitment and support to the employees and the organization itself. Setting a vision is fraught with pitfalls for the unwary. If it is a truism that the only certainty is uncertainty, and the only constancy is change, then leaders should expect a bumpy ride en route to their organization's place in the future. Organizations often face a future that is unclear. The competitive landscape (what rival firms are developing and planning to implement) adds to the complexity. Such situations, which are most of the time, may be exacerbated when an organization faces technological change or other disruptions to the way it likes to conduct its business operations.

Faced with an opaque roadmap of their future competitive landscape, organizational leaders may have three possible responses. A first (perhaps natural) response is denial. This may be compared to the head-in-the-sand approach taken by ostriches and other species when facing threats. Underpinning this approach tends to be a rationale that says: "Our organization is special, we have attributes that makes us unique in our industry, and we are immune to any external environmental changes that may affect other firms." This is tantamount to the statement that such and such an institution is too big to fail.[9] Perhaps an alternative optimistic statement was that the RMS *Titanic* was unsinkable.[10] A second response is resistance. This can take many forms, some of these being to find ways to try and prevent the elements of change from taking hold. Among the seemingly most popular is recourse to legal actions such as lawsuits and legal statutes. A third response is the most challenging, that is, to learn about the benefits and difficulties of implementing the change. In some cases this may require an organization to discard time-honored practices and processes. Some practices may have become entrenched in an organization's DNA and constitute the sole way of conducting operations. Leaders may thus need to make difficult choices: either to ensure that these practices and processes are recognized as critical success factors (CSFs) through which the organization has earned its reputation or, if these processes and practices have over time become generic, to develop new ways of working. Ideally, the organization develops new ways of working that cannot be readily replicated by other firms in an industry (i.e., cannot become generic and common to competitors).

As mentioned, a critical challenge for organizational leaders is to make sense of changing business landscapes brought about by emerging socio-economic developments and disruptive technologies. The future shape and format of our societies is not always clear from a current-day perspective. The future has the capacity to surprise. Organizational leaders need to heed the caveat given with financial products: past performance is not necessarily an indicator of future performance.

Leadership the Richard Branson Way

Richard Branson is the founder of the Virgin Group. This conglomerate now has over 400 branded businesses in its stable, including Virgin Records, Virgin Atlantic, and Virgin Galactic. Even as an early teenager, Branson was a serial entrepreneur.[11] This was evident from his early years at Stowe, the English public fee-paying boarding school that he entered at the age of thirteen. Richard and a neighborhood friend decided they could make money from selling Christmas trees. They made their decision during their Easter vacation from school, so there seemed to be sufficient time for the fresh plantings to grow into saleable products. Thus, they bought seedlings for £5 and calculated their healthy return on this initial investment over the Christmas season later that same year. Their Easter vacation over, both boys returned to their different boarding schools. Nature was allowed to take its course. Unfortunately, the course of nature also involved hungry rabbits whose healthy appetites decimated the young trees. Plan B for the young entrepreneurs was now crystal clear: they shot and skinned rabbits and sold these to a local butcher. While honor was no doubt satisfied, this did not generate the healthy revenues expected from a large batch of Christmas trees. As with all entrepreneurs, lack of success in their venture was seen as a mere glitch. The following year the boys evolved a second venture: breeding budgerigars. As the birds were bred in cages, hungry rabbits were not an issue.

The two entrepreneurs now faced a different problem. This was not supply (as had been the case with the Christmas trees); it was now demand. The budgerigars bred rapidly (almost like rabbits, you could say). If every potential customer in the local area bought two budgerigars, the young entrepreneurs would still have a glut of the birds on their hands. This standard business question of matching demand with supply was resolved when Richard received a letter from his mother to say that rats had gnawed into the birdcages and eaten all of the birds. In fact, Richard's mother had opened the cage doors to release the birds, as she was tired of caring for a constantly growing chatter of budgerigars (in both senses of the word). The next Branson venture was an events listings magazine. Called *Student*, the magazine was sold in universities, colleges, and to passersby on the street. By all accounts, this became a cash cow (and literally cash-driven). The magazine funded a student advisory center and then a record shop that ultimately became Virgin Records.[12] These relatively modest beginnings made Richard Branson a millionaire by the age of twenty-four.[13] Through constant division and multiplication of his enterprises, Branson has developed the single-branded, multicompanied conglomerate that has made him one of the richest people in the world.[14] This track record seems not too bad for someone who is dyslexic and found school work difficult.[15]

Leaders and Organizational Development

The famed Italian Renaissance polymath Michelangelo (1475–1574) is alleged to have described his relationship to his art by saying that he envisaged the beautiful statue imprisoned within every block of marble. His artistic genius was to be able to identify the contours of the eventual figure and sculpt the marble to free it so that others could see the beauty as he himself saw it imprisoned within the raw block of marble. This seems to convey the essence of organizational development. For organizational leaders the beauty of their as yet imprisoned organization may not always be as obvious as the hidden statues were for Michelangelo.

But, as with the art of sculpture, developing organizations toward service excellence requires persistence and patience on the part of leaders and their executives. It is not sufficient to say that employees are being encouraged to develop a service mind. A helpful starting point is to develop employees' sensitivity to customer wants and needs (i.e., focusing on service demand rather than service supply). Corporate statements about developing a service mind seem to ignore the important role of the customer, and particularly the customer's important contribution toward developing the desired service in collaboration with service-providing employees. In this way there is a convergence between the demand and supply strands of a service encounter.

At their heart, organizations comprise collections of people gathered together (ideally) for a common purpose. In developing their organizations for effective performance, leaders have a number of areas on which to focus. An effective leader takes a helicopter view and sees the landscape in which the organization conducts its business.

There are the levels of individual employees, teams (however defined and configured), and the whole organization. At each of these areas of focus, there are similar, overlapping, and diverse tasks for the organizational leaders.

At the level of individual employees the leader's tasks may relate to setting an environment where errors are tolerated (as long as these are used for learning). This may involve eradicating a blame culture where leaders and their managerial teams seek culprits and scapegoats every time a situation does not deliver anticipated results. The precarious nature of service can generate error and possible customer complaints or an outcome that is less than satisfactory for both the service user and the service provider. Developing an organizational environment in which employees feel valued and supported in their work efforts may not always be easy, but tends to be well worth a leader's efforts.

Leadership and Moments of Truth: Making a Difference in Service Organizations

Mahatma Gandhi (1869–1948) said that individuals should themselves be the change they wish to see. Leaders need to be seen as role models for their organizations, behaving as an exemplar of what is expected of other employees. This gives another reason why executives and managers in service organizations need to internalize concepts of service. It is important that executives and their managers do not view the service component of their organization as solely the responsibility of customer-facing employees. A critical success factor in a service organization is the ability of executives and senior management to internalize service and act on their knowledge by setting initiatives and empowering employees to act. In this way, leaders can embed the service agendas in their organization and set employees (at all levels) in the same direction toward service excellence.

Internalizing service applies to all employees, not solely those whose daily work routines involve direct contact with customers. By internalizing service, we mean understanding the very core of service: this involves understanding not only the what and the how of service delivery (but the who, why, and when of service). Together and separately we have worked in organizations that not only didn't know what their customers needed (which we would include in basic knowledge), but also, in some cases, did not have a clear picture of who their customers were. Conventionally, leaders have tended to ensure that frontline service-providing employees are tasked (and sometimes fully trained) to give service. The rationale seems to be that as these employees are the ones that the customer sees and from whom customers expect to receive their service, other employees do not need to be fully informed or educated in attributes of service provision. However, as we know from experience, service quality can be deficient because of shortcomings in service provided by support employees (such as units devoted to IT, procurement, budgeting, or human resources). These specialist parts of organizations hold key roles in providing relevant services to *internal* customers.

As we outlined in Chapter 1, a service encounter concept covers the provision of customers' needs and also the component contributory features of an organization's internal processes that support the delivery of the customers' needs through the intermediary of service-providing employee(s). In essence, these components are the what and the how of service provision that we earlier categorized as basic knowledge. The service encounter thus requires not only service provision (satisfying customers' needs), but also facilities within the organization that ensure that customers' needs will be satisfied time after time and to a consistent level of quality.

Behind the service-providing employees lies "the chain of local and central activities needed to produce the service."[16] The military calls this action and supply relationship the teeth-to-tail ratio. In former times, armies needed more teeth (armed forces facing the enemy) than tail (support services of logistics and supply). Over time, as armies relied less on handheld weapons (bows and arrows, halberds, muskets) and more on increasingly heavy weaponry (trebuchets, siege engines, cannons, and tanks), the tail of logistics needed to equip, support, and service the front line grew longer and less manageable. In 2003, Private First Class Jessica Lynch was among U.S. soldiers captured when her supply convoy was ambushed near the town of Nasiriyah in southern Iraq.[17] A lengthy tail of supply is said to have partly contributed to Emperor Napoleon's defeat in his Russian campaign of 1812, leading to the tragic retreat of his army from Moscow.[18]

In the provision of service, the importance of so-called back office support services cannot be overemphasized. Although generally out of sight of the customer, without the support of these personnel, the service provision could not wholly function. Hence, an organization's leaders need to ensure that responsibility for service delivery is part of their organization's ethos.

In delivering service, an organization must get right to the basics, that is, recruiting and then training competent people to occupy customer-facing roles. Competent management of frontline service encounters is a necessary but not sufficient condition for serviced success. Supporting the frontline service providers are the employees tasked to maintain the background needs for service delivery. It is therefore critical that a service organization successfully manages its frontline service encounters. In the short term customer satisfaction is at stake. In the longer term, the perceived reputation of the organization is at stake. Continued success (or otherwise) of the service encounters will likely determine the survival of the organization.

Scandinavian Airlines System (SAS), during the era of Jan Carlzon's presidency (from 1981 to 1993), was, as far as we are aware, the first international company to integrate on a large scale the concepts of moments of truth as key foundations of its business growth and strategic development through service management. Under Carlzon's leadership SAS regarded its employees as crucial to business success. Famously, he redefined the business tasks of all employees so that "the entire company—from the executive suite to the most remote check-in terminal—was focused on service."[19] Jan Carlzon was invited to be SAS president when the airline was in a crisis. In the previous two years the airline had posted losses of US$30 million and was rated near the bottom of the European airlines for its lack of punctuality. Within a year of Carlzon's arrival, SAS had returned to profit. By 1984 SAS was voted Air Transport World's "Airline of the Year."[20] During his first years as president, Carlzon initiated 147 projects to improve customer service. With a strategic focus and business emphasis on service, an integral part of the airline's strategic development became development of human resources (HR). In this context HR incorporated development of skills and competences to transform SAS into a service-oriented airline. Three hundred sixty degree evaluations become a yearly process throughout the entire group of companies. In the process Jan Carlzon became a world-renowned management guru (*Moments of Truth* became a best seller). In his introduction to the book, Tom Peters writes: "Carlzon charged the frontline people with 'providing the service they had wanted to provide all along.'"[21]

The important role played by frontline employees in a service encounter is shown in Figure 6.1. We have interpreted this figure from its original use with service in a manufacturing organization to service given face-to-face.

As the figure shows, the employee contributes several and varied types of knowledge and skills to the encounter. In addition, the employee is ideally positioned to receive direct (i.e., firsthand) feedback from the customer. The important role played by the employee is indicated in the figure by the attributes "control over resources," "self-scheduling," "personal accountability," and "direct communications authority." Hence, the employee "possesses" features that become integral parts of the service encounter. Perhaps most importantly, the employee is in a position not only to audit the customer's response (through feedback, both formal and informal), but also to assess the organization's readiness to deliver attributes of the service. For example, the employee has control of the organization's resources for delivering service. These are likely to include human resources (any other employees involved in the service delivery, such as the concierge, receptionists, and bellhops when a guest arrives at a hotel), fixed assets (such as equipment),

FIGURE 6.1
An employee's relationships with the customer. (From Frederick Herzberg (1989), Motivation and Innovation: Who Are Workers Serving? *California Management Review*, 22(2), 61.)

financial resources (such as taking fees, giving change, and swiping credit cards), and environmental resources (such as the market space where the service encounter occurs). In part of the service encounter the employee controls the scheduling of himself or herself and others involved in the service delivery. If the scheduling is uncoordinated or just plain wrong, then the service delivery begins to unravel.

This is especially so when the scheduling relates to timing (particularly sequentially timed parts of the service) and the content of the service offering. Service users tend to become upset or irritated when parts of a service are out of their intended or logical sequence (for example, if the restaurant bill arrives before coffee has been served). The employee's responsibility for personal accountability is unavoidable in a face-to-face service encounter. Thus, the emotional stakes are high for both the service user and the employee. When service is delivered in a face-to-face setting, there are very few ways in which the emotional content of the encounter can be avoided. For some commentators, emotionality is at the core of service delivery and service use.[22]

In any organization, unhappy employees tend to create an unhappy, often dysfunctional, workplace. Naturally, this has an adverse effect on employee morale. Although initially, low levels of morale relate to only a few employees, the feeling can be contagious. Before too long, low motivation can pervade a whole department and then the whole workplace. In an organization oriented toward service, low morale is a threat to the service ethos. Low morale can cross the border between inside the organization (employees) and outside the organization (customers), and thus influence aspects of the moment of truth.

The up close and personal nature of service means that customers sense when employees are unhappy, partly because customers often feel the impact of employees' moods. Who hasn't experienced a poor service attitude by a service provider who is obviously in a bad mood? Service settings, in which customers and employees interact at the moments of truth, have an added importance in service quality when employees' moods form part of the customer's experience. Organizational climate (in essence, whether or not the organization is a good place in which to work) is an influencing factor on how employees approach their work tasks and their job overall.[23] Organizations tend to be more civilized if executives and other employees in positions of power or even colleagues don't display obnoxious behaviors that sully the workplace environment.[24] To ensure that the organization doesn't tolerate antisocial behaviors (which may sometimes border on the psychopathological), rules are needed to identify and deal with the employees who display these behaviors to the detriment of the workplace environment.[25] When abhorrent behaviors are discernable in some employees, an organization's leaders need to take steps to eradicate these behaviors. Left unchecked, dysfunctional behaviors may become the norm rather than the exception, and thereby exert an undue influence on the organization's culture.[26]

Leaders (and leaders-in-waiting) would be advised to reflect on the following quotation: "The race is not to the swift, nor the battle to the strong."[27]

Endnotes

1. See Robert J. Allio (2013), Leaders and Leadership: Many Theories but What Advice Is Reliable? *Strategy and Leadership*, 41(1), 4–14.
2. See Richard A. Barker (2001), The Nature of Leadership, *Human Relations*, 54(4), 469–494; Robert J. Allio (2005), Leadership Development: Teaching versus Learning, *Management Decision*, 43(7–8), 1071–1077.
3. Robert Goffee and Gareth Jones (2000), Why Should Anyone Be Led by You? *Harvard Business Review*, September-October, pp. 63–70; Simon Sinek (2014), *Leaders Eat Last: Why Some Teams Pull Together and Others Don't*, London: Penguin Books.
4. Robert Goffee and Gareth Jones (2005), Managing Authenticity: The Paradox of Great Leadership, *Harvard Business Review*, December, pp. 87–94.
5. See various discussions in Peter M. Senge (1990), The Leader's New Work: Building Learning Organizations, *Sloan Management Review*, 32(1), 7–23; Richard A. Barker (2001), The Nature of Leadership, *Human Relations*, 54(4), 469–494; Robert J. Allio (2005), Leadership Development: Teaching versus Learning, *Management Decision*, 43(7–8), 1071–1077; Roger L. Martin (2009), *The Design of Business: Why Design Thinking Is the Next Competitive Advantage*, Boston: Harvard Business School Press; Seth Kahan (2010), *Getting Change Right: How Leaders Transform Organizations from the Inside Out*, San Francisco: Jossey-Bass; John R.

Latham (2013), A Framework for Leading the Transformation to Performance Excellence Part II: CEO Perspectives on Leadership Behaviors, Individual Leader Characteristics, and Organizational Culture, *Quality Management Journal*, 20(3), 19–40; Robert J. Allio (2013), Leaders and Leadership: Many Theories but What Advice Is Reliable? *Strategy and Leadership*, 41(1), 4–14.

6. See the discussions in John P. Kotter and James L. Heskett (1992), *Corporate Culture and Performance*, New York: Free Press, especially Chapters 2, 3, and 6.

7. John Kotter and James L. Heskett (1992), *Corporate Culture and Performance*, New York: Free Press.

8. See Edgar H. Schein (1999), *The Corporate Culture Survival Guide: Sense and Nonsense about Culture Change*, Jossey-Bass, Wiley Publishers; Jerome H. Want (2007), *Corporate Culture: Illuminating the Black Hole*, New York: St. Martin's Press; Terrence E. Deal and Allan A. Kennedy (2000), *Corporate Cultures: The Rites and Rituals of Corporate Life* (rev. ed.), New York: Perseus Books.

9. See Alan Ross Sorkin (2009), *Too Big to Fail: The Inside Story of How Wall Street and Washington Fought to Save the Financial System—and Themselves*, New York: Viking Press. The book describes the 2008 financial crisis and the collapse of Lehman Brothers (the institution that was said to be too big to fail).

10. The RMS *Titanic* hit an iceberg and sank on April 15, 1912, with a loss of over 1,500 lives. The vessel was on her maiden voyage from Southampton to New York. Looked at realistically and admittedly with 20:20 hindsight, the *Titanic* was a ship and, in common with other ships, if the hull is punctured below the waterline, she sinks.

11. See Richard Branson (1998), *Losing My Virginity: The Autobiography*, London: Virgin Publishing, especially pp. 33–70.

12. For various aspects of Richard Branson and the Virgin Group, see Tim Jackson (1998), *Virgin King: Inside Richard Branson's Business Empire*, London: Harper Collins. Also see Manfred F.R. Kets De Vries (1998), Charisma in Action: The Transformational Abilities of Virgin's Richard Branson and ABB's Percy Barnevek, *Organizational Dynamics*, Winter, 7–21; Manfred F.R. Kets De Vries and Elizabeth Florent-Treacy (1999), *The New Global Leaders: Richard Branson, Percy Barnevek and David Simon*, San Francisco: Jossey-Bass.

13. See Des Dearlove (2007), *Business the Richard Branson Way: 10 Secrets of the World's Greatest Brand Builder* (3rd ed.), Chichester, UK: Capstone Publishing.

14. See Robert M. Grant (2004), *Cases in Contemporary Strategy Analysis*, Malden, MA: Blackwell Publishers. Richard Branson and the Virgin Group of Companies in 2004.

15. For analyses of Richard Branson's character and personality, see Larisa V. Sahvinina (2006), Micro-Social Factors in the Development of Entrepreneurial Giftedness: The Case of Richard Branson, *High Ability Studies*, 17(2), 225–235; Todd A. Finkle (2011), Richard Branson and Virgin, Inc., *Journal of the International Academy of Case Studies*, 17(5), 109–121.

16. Bo Edvardsson (1997), Quality in New Service Development: Key Concepts and a Frame of Reference, *International Journal of Production Economics*, 52(1–2), 31–46.

17. Rick Bragg (2003), *I Am a Soldier Too: The Jessica Lynch Story*, New York: Random House Books.

18. Richard A. Hardemon (2011), General Logistics Paradigm: A Study of the Logistics of Alexander, Napoleon and Sherman, *Air Force Journal of Logistics*, 35(1–2), 81–83. Also see Mark J. Kroll, Leslie A. Tombs, and Peter Wright (2000), Napoleon's Tragic March Home from Moscow: Lessons in Hubris, *Academy of Management Executive*, 14(1), 117–128.

19. Jan Carlzon (1987), *Moments of Truth*, Cambridge, MA: Ballinger Publishing, p. 26.

20. Management guru Tom Peters wrote the introduction to *Moments of Truth* (this quotation is from p. ix). Tom Peters and Robert Waterman were co-authors of *In Search of Excellence* (Warner Books: NY, 1984). At the time of writing this classic management book, both authors worked at the consultancy group McKinsey & Company.

21. Tom Peters in his introduction to *Moments of Truth* (p. x).

22. See, for example, Arlie Hochschild (1983), *The Managed Heart: Commercialization of Human Feeling*, Berkeley: University of California Press; Stephen Fineman (ed.) (2000), *Emotion in Organizations*, London: Sage Publications; Marek Korczynski (2003), Communities of Coping: Collective Emotional Labour in Service Work, *Organization*, 10, 55–79; Steven Henry Lopez (2010), Workers, Managers, and Customers: Triangles of Power in Work Communities, *Work and Occupations*, 37(3), 251–271; Janine L. Bowen (2014), Emotion in Organizations: Resources for Business Educators, *Journal of Management Education*, 38(1), 114–142.

23. See discussions in Manfred Kets de Vries (2001), Creating Authentizotic Organizations: Well-Functioning Individuals in Vibrant Companies, *Human Relations*, 54(1), 101–111; Russell Cropanzano and Thomas A. Wright (2001), When a "Happy" Worker Is Really a "Productive" Worker: A Review and Further Refinement of the Happy-Productive Worker Thesis, *Consulting Psychology Journal: Practice and Research*, 53(3), 182–199; Thomas A. Wright, Russell Cropanzano, Philip J. Denney, and Gary L. Moline (2002), When a Happy Worker Is a Productive Worker: A Preliminary Examination of Three Models, *Canadian Journal of Behavioural Science/Revue Canadienne des Sciences du Comportement*, 34(3), 146–150; Armenio Rego and Miguel Pina e Cunha (2008), Authentizotic Climates and Employee Happiness: Pathways to Individual Performance? *Journal of Business Research*, 61(7), 739–752.

24. Interesting discussions of the darker side of corporate life can be found in Manfred Kets de Vries and Danny Miller (1984), *The Neurotic Organization: Diagnosing and Changing Counterproductive Styles of Management*, San Francisco: Jossey-Bass; Christine Clements and John B. Washbush (1999), The Two Faces of Leadership: Considering the Dark Side of Leader-Follower Dynamics, *Journal of Workplace Learning*, 11(5), 170–175; Peter J. Frost (2003), *Toxic Emotions at Work*, Boston: Harvard Business School Press; Christine M. Pearson and Christine L. Porath (2005), On the Nature, Consequences, and Remedies of Workplace Incivility: No Time for 'Nice'? Think Again, *Academy of Management Executive*, 19(1), 7–18; Manfred Kets de Vries (2006), The Spirit of Despotism: Understanding the Tyrant Within, *Human Relations*, 59(2), 195–220; Roderick M. Kramer (2006), The Great Intimidators, *Harvard Business Review*, February, pp. 88–97; Robert Sutton (2007), *The No Asshole Rule: Building a Civilized Workplace and Surviving One That Isn't*, London: Sphere Books.

25. See Robert Sutton (2004), More Trouble Than They're Worth, *Harvard Business Review*, 82(2), 19–20; Robert Sutton (2007), *The No Asshole Rule: Building a Civilized Workplace and Surviving One That Isn't*, London: Sphere Books. See also

Kate Ludeman and Eddie Erlandson (2004), Coaching the Alpha Male, *Harvard Business Review*, 82(5), 58–67; Kate Ludeman and Eddie Erlandson (2006), *Alpha Male Syndrome*, Boston: Harvard Business School Press.

26. For suggestions for dealing with dysfunctional employees, see Robert Sutton (2007), *The No Asshole Rule: Building a Civilized Workplace and Surviving One That Isn't*, London: Sphere Books.

27. Ecclesiastes 9:11.

7

Leading Organizations as if People Matter: Humanist Approaches

Leadership Matters and People Matter

Leadership matters. Napoleon Bonaparte (1769–1821) famously demonstrated empathy with his rank-and-file soldiery when he asserted that "an army marches on its stomach." An army needs to have its members at the utmost state of readiness, whether physical, mental, or in terms of equipment. This is especially so when the army is set to do battle (as was often the case for the French army under the leadership of Napoleon). Soldiers whose diet is deficient are unlikely to be at peak fitness to engage in combat.

Leadership matters, especially to employees. Astute leaders know that employee well-being is an essential precondition to generating high levels of motivation required to do their work tasks to any level of quality. Psychological capital (known as PsyCap), the development of positive human traits in the workplace, is predicted to have a positive influence on desired outcomes such as increased performance.[1] Positive workplace environments are acknowledged to contribute to employee confidence, hope, optimism, happiness, and emotional intelligence (making the acronym CHOSE).[2] The leader's tasks include ensuring that employees work in a positive environment.[3] Even the provision of on-site refectories or cafeterias may make a difference in employee motivation. Singly and together we have worked in institutions where on-site eating arrangements were poor. In one institution that we know well the food service facility is tolerated by employees, as the location is convenient and the food is cheap (albeit limited in cuisine). And, regardless of complaints to senior executives, the situation has not changed. This has given rise to several knock-on effects. First, employees regard their senior managers as insensitive and uncaring about their welfare and seek out other examples from their daily work to confirm their grievances. Second, employees take longer lunch breaks, for example, to travel to nearby shopping centers where the choice of eating places is diverse. Extending the conventional lunch break period is regarded as a way to get back at their institution. Third, employees unfavorably compare their own working conditions with other organizations and competitors that they know. One situation has thus become a touchstone for a wider range of employee-related grousing. Such was the depth of feeling that this has become a talking point with new and existing employees. New employees bring their recent experiences to the discussion. Employee ill feeling can easily spread through an organization, a sort of oil spot effect when a small stain spreads slowly and seeps into the fabric. Ill feeling damages employee motivation in the short term, and in the longer term may harm the reputation of the organization. If there is any positive effect, it is that such a situation can be interpreted as a means whereby colleagues reinforce their solidarity. However, this may in turn open and then reinforce a schism between senior managers and employees (a cultural perception of them and us). Fourth, managers make a point of taking visitors to lunch elsewhere and explaining to them that the in-house eating facilities are unsuitable.

Over time, several people within the organization took jobs in a competitor institution. They then realized how poorly their personal needs had been served in the former institution. A month or two after taking up their new posts, and asked about work conditions at their new institution, one person said, "Wow! We couldn't believe it. They have three choices of menu, Thai, Indian, and Western, and they change the menu every day. Friday is a banquet lunch. This is enough to get me out of bed in the mornings." As is frequently the case in organizations, the concerns of executives and senior managers, such as grand strategy, building international alliances, gaining market share, pricing strategy, and structures, are a world away from the everyday concerns of employees. It is likely that addressing these everyday concerns, such as refectory food, comfortable workspaces, and amenable colleagues, would give the executives and managers a more committed workforce willing to dedicate themselves to addressing and delivering the big picture concerns. When the world outside the workplace is increasingly characterized by uncertainty, turbulence, and persistent change, employees feel that their employers should be caring.[4] Similarly, employees remunerated "on the cheap" are likely to become demoralized, especially if the pay scales are poor in comparison to those of competitors. As the popular saying in business and management has it: *you pay peanuts, you get monkeys.*

Employee commitment is defined as "an employee's identification with and agreement to pursue the company's or the unit's mission."[5] Competent leadership ensures that the organization has a workplace climate and environment that supports employees (at all levels) in their work. This is one task for an organization's leaders. Arguably, it is one of their most important tasks. It is also one of the more difficult tasks facing a leader. A number of authors and commentators advocate that there is a range of key activities on which organizational leaders should focus their attention and efforts. The suggested range is neither wide nor diverse, and there is (more or less) consensus of what these should be.

Leaders' tasks include setting a vision, building coalitions, developing a positive organization culture (sometimes called energizing), mobilizing employee commitment, developing employees (e.g., through empowerment), teams, and future leaders, aligning organizational goals with remunerations and incentives, and adapting the organization to changing environments.[6] In essence, leaders are tasked to focus on delivering ends (the expressed strategic goals of their organization) by mobilizing the available means. Ends include managing the organization's objectives and motivating the workforce by enabling individuals to achieve their own and their organization's ambitions.[7]

Competent leaders develop a vision for their organization and design relevant workable strategies to achieve this vision. Publicizing the vision through channels of communication is an essential element in gaining employee sign-up for future direction of the organization. The famed Italian Renaissance polymath Michelangelo (1475–1574) is alleged to have described his relationship to his art by saying that he envisaged the beautiful statue imprisoned within every block of marble. His artistic genius was to be able to identify the contours of the eventual figure and to sculpt the marble to free it so that others could see the beauty as he himself saw it in the raw block of marble.

The Humanistic School of Management

The leadership styles of Richard Branson, Vineet Nayar Ricardo Semler, and Ralph Stayer echo the philosophies of key management thinkers of the twentieth century. These include Mary Parker Follett (1868–1933),[8] Elton Mayo (1880–1949),[9] Abraham Maslow (1908–1970),[10] and Frederick Herzberg (1923–2000).[11] For each of these influential thinkers the theory and practice of management should incorporate the human side of work and the workplace. Although their separate focuses are diverse, their work has a common thread. In their own ways, each of these management thinkers perceived the workplace as essentially a social setting in which human relationships played a large part. For this reason, their work is known collectively as the humanistic school of management. By definition, humanism is "an outlook emphasizing common human needs and is concerned with human characteristics."[12]

The humanistic school of management thought was in marked contrast to theories about workplace efficiency (designing work around processes and machinery to optimize output) and workplace effectiveness (optimizing output by designing work around people). Mary Parker Follett's writings focused on the roles and functions of people in organizations, particularly people working in group settings and employees' involvement in decision making.[13] Her theories, which seemed to be ahead of her time, earned her the epithet the "prophet of management."[14] In her writings, Follett is notable for describing the benefits of what is now called empowerment. Her perspective on this topic is that group membership helps individual employees attain their true potential, both as people and as employees.[15] With group membership there is a symbiosis between the group and its constituent members: the group functions through its individual members, while each individual member gains self-esteem through being a member of the group. This is especially so when the group is recognized for its achievements.

Follett advocated that organizations should discard the dichotomy between leaders and employees (in essence, them and us), or constructed hierarchies. Instead, organizations are advised to move toward collaborative participatory practices that are better able to help employees reach their potential.[16] Elton Mayo is best known for a five-year series of experiments conducted from 1927 to 1932 with workers in the Hawthorne Electric Plant of Western Electric in Cicero, Illinois, near Chicago. Called the Hawthorne studies, the focus of these experiments was to analyze and assess changes in worker productivity in response to changes made to their work environment. Mayo and a team of co-researchers from Harvard University made changes to lighting (brighter or dimmer), the length of the lunchtime breaks, and the length of the working week.[17] Findings from the experiments can be summed up in the sentence: "A human problem requires a human solution."[18]

The humanistic approach to management studies in the workplace represented a gradual change of focus in management thinking. Early twentieth-century thinkers on business and management were F.W. Taylor (1856–1915), Frederique Fayol (1841–1925), and Henry Ford (1863–1947), whose work focused on work processes.[19] The overwhelming emphasis on well-functioning machinery to the detriment of the well-being of people was lampooned in Charlie Chaplin's 1936 classic silent film *Modern Times*. Chaplin's little tramp character is assigned a manual job tightening nuts on a factory production line. The speed of the line is increased on the orders of the boss. The work is repetitive and the little tramp tries to keep up his pace of work. Driven crazy by the routine, he slides along a conveyor belt and becomes parts of the cog wheels of the machine. The film is well worth watching for its critique of production line processes.

A key role for an organization's leaders involves communication. In this context communication is multidirectional and multichanneled. Communication needs to be effective in both form and function. Form to address relevant audiences and stakeholders, whether these are inside the organization (such as employees with various roles or tasks) or outside the organization (such as customers, suppliers, investors, and government representatives). Communication within the organization is crucial to its effective operations. Alignment of internal resources toward specific stated goals is, in the final analysis, the task of the organizational leadership. Relevant and effective communication helps ensure success. When communication is a weak link in an organization's processes, anomalies tend to occur. It is important to avoid the situation (undoubtedly apocryphal) in command and control economies where the timber yard continues to lathe handles for spades, oblivious to the fact that the steel mill no longer produces the metal blades to be affixed to the wooden handles to make the complete tool.

A key facet of internal communication is to state clearly a vision for direction and proposed actions for progressing from the present into the future. This is the essence of organizational strategy. While such a statement is essential for employees within an organization, it is also necessary to communicate the vision to external communities, such as customers, suppliers, investors, and the financial markets. Part of a vision statement will communicate expectations for workplace tasks and behaviors, especially standards of quality. It is worthwhile communicating this feature of the vision to the external communities, especially customers. Setting expectations for service quality for both employees and customers aids in the construction of shared (co-created) service. When customers contribute to service encounters, there is an increased possibility of customer high satisfaction.

Employees Come First

The Hawthorne studies conducted by Elton Mayo and his team of researchers suggested that employees were less concerned about pay and other financial remuneration and more concerned about being part of a workplace group. In the Mayo experiments the attention paid to the employees by the researchers encouraged increases in worker productivity.[20]

Leaders know they must build coalitions of employees in order to mobilize the available human resource competences to execute their strategies. In this, leaders need to initiate and direct the necessary changes to the organization to align the resources and ensure these are focused in the same direction. In all of these strands of organizational direction, successful leaders realize that communication is a critical component in their tool kit. Employees are a critical resource in a leader's quest to mobilize the means (resources) in order to deliver the projected ends (outputs and outcomes). Serial entrepreneur and billionaire Richard Branson says that in his companies his employees come first.[21] He is quite clear and emphatic in his rationale: "It just seems common sense to me that if you start off with a happy well motivated workforce, you're much more likely to have happy customers. And in due course the resulting profits will make your shareholders happy."[22] Another organization that seems to be thriving from a happy workforce is Zappos.com.[23] Zappos is an online shoe and accessory retailer that is now worth over $1 billion. Zappos CEO Tony Hseih sold his earlier Internet venture to Microsoft because he said he didn't enjoy working in the culture that it had become. In 1999 he invested in an Internet start-up company called ShoeSite.com, later renamed Zappos. In 2009 Amazon.com bought Zappos, but the message from Amazon CEO Jeff Bezos was to continue to have the Zappos culture happiness of employees and "customer obsession."[24] Happiness in the workplace has positive effects on both employee productivity and reducing stress at work.[25] As Ricardo Semler, CEO of Semco, based in Brazil, says: "Clearly, workers who control their working conditions are going to be happier than workers who don't."[26]

When Ralph Stayer took over his family's sausage processing business in Sheboygan Falls, Wisconsin, the company was growing, generating healthy revenues and returning profits. However, Stayer was still dissatisfied.[27] The workforce was lackadaisical and seemed to be going through the motions of doing their jobs. And unlike many CEOs, he considered that any deficiencies in his organization were due to his own failing as a leader. Stayer reflected on the business founded by his parents in 1945. He decided an approach to his unease was a lack of focus on developing the employees. As Stayer wrote in his case study, "People want to be great. If they aren't, it's because management won't let them be."[28]

Richard Branson focuses on his employees. He has stated his business philosophy as putting his employees first and his customers second. Investors are third in order of priority.[29] This ordering somewhat reverses conventional business wisdom, which advocates that the customer is king and the customer is always right. The rationale behind an employees-first policy is that unless the service providers are given relevant training, and feel they are well treated and compensated by their employer, they can hardly do a good job of looking after customers. Hence, astute companies such as Semco and the Virgin Group ensure that employees are a top priority for executives and their managers. The focus on employees by an organization's leader is not simply altruism or kindheartedness. It also makes sound business sense. Doug Conant, CEO of Campbell Soup, says that his employees receive training and development as part of the Campbell promise of valuing people.[30] The Campbell brand is certainly iconic in part thanks to the pop art of Andy Warhol (1928–1987), whose silk screen images of the Campbell tomato soup can in the 1960s became an instantly recognizable feature of the pop art movement. And founded in 1869, the Campbell Soup company has enjoyed more than fifteen minutes of fame.[31] CEO Doug Conant describes a six-stage process of leadership in the company: inspire trust, create direction, drive organization alignment, build organization vitality, execute with excellence, and produce extraordinary results.[32]

In organizations, if employees themselves are not content in their workplace (perhaps because they feel treated as second-class citizens by executives, managers, or other employees), how can they perform in an upbeat manner in front of customers? Vineet Nayar, the CEO of HCL Technologies, took a creative approach in engaging his employees.[33] Nayar's approach included sharing financial data (previously held as confidential and for executives' eyes only), posting all employee appraisals (including that of the CEO) on the company's intranet, and taking employees to visit customers to resolve problems. The intention of these and other surprising initiatives was to promote transparency and engage employees. Another surprising feature at HCL Technologies is that any employee can give feedback to anyone else in the company. Predictably, the CEO's e-mail inbox overflows with employee feedback. These initiatives influenced the behaviors of HCL employees toward their customers. In HCL Technologies, the interface between customers and employees is called the value zone.[34]

Brazilian entrepreneur Ricardo Semler would find a kindred spirit in British entrepreneur Richard Branson. Each leads his business group with a focus on his employees. In 1980, when he was twenty-one years of age, Ricardo Semler was appointed CEO of the business begun by his father.[35] Originally called Semler and Company, and then renamed Semco SA, the company is located in the outer suburbs of São Paulo, Brazil. The company manufactures industrial equipment such as oil pumps for the petroleum industry and kitchen equipment, sold mainly to restaurants. Ricardo had joined the company two years earlier with the grandiose job title of assistant to the board of directors. This was a job in name alone, as the duties were few and far between. In his spare time Semler was playing in a rock band. Needless to say, Ricardo became bored at the office. When his father retired, he appointed Ricardo as CEO so that he could make his mistakes while his father was still alive.[36] Ricardo made radical changes, including firing 30 percent of his father's senior managers and introducing the seven-day weekend (a shift system that gives employees seven days away from the workplace).[37] Semco has three corporate values: democracy (which means employee participation), profit sharing, and information (transparency).[38] Employees work the hours they choose within a flex-time schedule. The company has no organizational structure, and therefore no organization chart, and no business or strategic plan, and has done away with goals and mission statements, as well as long-term budgeting.[39] The company has an unorthodox style of governance (there isn't one), but has seen huge growth with increased profits, with employees who are highly motivated, with a resultant low turnover.[40]

An employee's feelings about his or her employer influences how that employee interacts with his or her customer. There is a recognized direct link between attributes of leadership, the organizational climate within the unit that provides service to customers, the environment in which the service is provided, and (ultimately) customer satisfaction with the service encounter.[41] Positive customer satisfaction leads to sales (and often, repeat customers).[42]

Endnotes

1. See Fred Luthans and Carolyn M. Youssef (2007), Emerging Positive Organizational Behavior, *Journal of Management*, 33(3), 321–349.

2. Fred Luhans (2002), Positive Organizational Behavior: Developing and Managing Psychological Strengths, *Academy of Management Executive*, 16(1), 57–72.
3. See, for example, Fred Luthans, Carolyn M. Youssef, David S. Sweetman, and Peter D. Harms (2013), Meeting the Leadership Challenge of Employee Well-Being through Relationship PsyCap and Health PsyCap, *Journal of Leadership and Organizational Studies*, 20(1), 118–133.
4. See Jason M. Kanov, Sally Maitlis, Monica C. Worline, Jane E. Dutton, Peter J. Frost, and Jacoba M. Lilius (2004), Compassion in Organizational Life, *American Behavioral Scientist*, 47(6), 808–827; Sandra M. Wilson and Shann R. Ferch (2005), Enhancing Resilience in the Workplace through the Practice of Caring Relationships, *Organizational Development Journal*, 23(4), 45–60; Fred Luthans and Carolyn M. Youssef (2007), Emerging Positive Organizational Behavior, *Journal of Management*, 33(3), 321–349; Michael Kroth and Carolyn Keeler (2009), Caring as a Managerial Strategy, *Human Resource Development Review*, 8(4), 506–531.
5. Gary Dressler (1999), How to Earn your Employees' Commitment, *Academy of Management Executive*, 13(2), 58.
6. See Ian Brooks (1997), Leadership of a Cultural Change Process, *Health Manpower Management*, 23(4), 113–119; Kets de Vries (1994), *Leadership Mystique Academy of Management Perspectives*, 8(3), 73–79; Ron Cacioppe (2000), Creating Spirit at Work: Re-Visioning Organizational Development and Leadership Part I, *Leadership and Organizational Development Journal*, 21(1–2), 48–54; Ron Cacioppe (2000), Creating Spirit at Work: Re-Visioning Organizational Development and Leadership Part II, *Leadership and Organizational Development Journal*, 21(1–2), 110–119; Robert S. Kaplan and David P. Norton (2004), Measuring the Strategic Readiness of Intangible Assets, *Harvard Business Review*, 82(2), 52–63; Vineet Nayar (2010), *Employees First, Customers Second: Turning Conventional Management Upside Down*, Boston: Harvard Business School Publishing.
7. See Julian Birkinshaw and Jules Goddard (2009), What Is Your Management Model? *MIT Sloan Management Review*, 50(2), 81–91.
8. See Dafna Eylon (1998), Understanding Empowerment and Resolving Its Paradox: Lessons from Mary Parker Follett, *Journal of Management History*, 4(1), 16–28; David M. Boje and Grace Ann Rosile (2001), Where's the Power in Empowerment? Answers from Follett and Clegg, *Journal of Applied Behavioral Science*, 37(1), 90–117.
9. Elton Mayo (1930), The Human Effect of Mechanization, *American Economic Review*, March, pp. 156–176; Elton Mayo (2003 [1933]), *The Human Problems of an Industrial Civilization*, Abingdon, Oxon, UK: Routledge, especially Sections III and IV.
10. Abraham Maslow (1943), A Theory of Human Motivation, *Psychological Review*, 50(4), 370–396; Abraham Maslow (1987 [1954]), *Motivation and Personality* (3rd ed.), New York: Addison Wesley Longman; Abraham Maslow (1998), *Maslow on Management*, New York: John Wiley & Sons.
11. Frederick Herzberg, Bernard Mausner, and Barbara Bloch Snyderman (2010 [1959]), *The Motivation to Work*, New York: Wiley; Frederick Herzberg (1965), The New Industrial Psychology, *Industrial and Labor Relations Review*, April, pp. 364–376; Frederick Herzberg (1966), *Work and the Nature of Man*, New York: World Publishing Company; Frederick Herzberg (1974), Motivation-Hygiene Profiles, *Organizational Dynamics*, September, pp. 18–29.

12. Domènec Melé (2003), The Challenge of Humanistic Management, *Journal of Business Ethics*, 44, 78.
13. See various chapters by Mary Parker Follett in Mary Godwyn and Jody Hoffer Gittell (eds.) (2011), *Sociology of Organizations: Structures and Relationships*, Thousand Oaks, CA: Pine Forge Press (imprint of Sage Publications).
14. Pauline Graham (ed.) (2003), *Mary Parker Follett Prophet of Management: A Celebration of Writings from the 1920s*, Washington, DC: Beard Books (by arrangement with Harvard Business School Press).
15. See Dafna Eylon (1998), Understanding Empowerment and Resolving Its Paradox: Lessons from Mary Parker Follett, *Journal of Management History*, 4(1), 19.
16. See relevant discussions in Dafna Eylon (1998), Understanding Empowerment and Resolving Its Paradox: Lessons from Mary Parker Follett, *Journal of Management History*, 4(1), 16–28; David M. Boje and Grace Ann Rosile (2001), Where's the Power in Empowerment? Answers from Follett and Clegg, *Journal of Applied Behavioral Science*, 37(1), 90–117.
17. For details see Richard Stillman II, Internal Dynamics: The Concept of the Informal Group, in *Public Administration: Concepts and Cases* (9th ed.), Boston: Wadsworth Cengage Learning, Chapter 6.
18. Richard Stillman II, Internal Dynamics: The Concept of the Informal Group, in *Public Administration: Concepts and Cases* (9th ed.), Boston: Wadsworth Cengage Learning, p. 147.
19. See discussions in Domènec Melé (2003), The Challenge of Humanistic Management, *Journal of Business Ethics*, 44, 77–88.
20. For details see Richard Stillman II, Internal Dynamics: The Concept of the Informal Group, in *Public Administration: Concepts and Cases* (9th ed.), Boston: Wadsworth Cengage Learning, Chapter 6. Also, Elton Mayo (1930), The Human Effect of Mechanization, *American Economic Review*, March, pp. 156–176; Elton Mayo (2003 [1933]), *The Human Problems of an Industrial Civilization*, Abingdon, Oxon, UK: Routledge, especially Sections III and IV.
21. Todd A. Finkle (2011), Richard Branson and Virgin, Inc., *Journal of the International Academy for Case Studies*, 17(5), 109–121, especially pp. 117–118.
22. Sue Minton (2010), Managing Human Resources in the Leisure Industry, *Managing Leisure*, 15(1–2), 1–3.
23. See Tony Hseih (2010), *Delivering Happiness: A Path to Profits, Passion and Purpose*, New York: Hachette Books; Anne Perschel (2010), Work-Life Flow: How Individuals, Zappos, and Other Innovative Companies Achieve High Engagement, *Global Business and Organizational Excellence*, July–August, pp. 17–30; Marguerite McNeal (2013), A Case for Culture: Why Aligning Employee and Company Values around 'Delivering Happiness' Matters, *Marketing Insights*, Fall, pp. 44–45.
24. See Deborah L. Cowles, Jan P. Owens, and Kristen L. Walker (2013), Ensuring a Good Fit: Fortifying Zappos' Customer Service and User Experience, *International Journal of Marketing Communications*, Fall, pp. 57–66.
25. Joanne H. Gavin and Richard O. Mason (2004), The Virtuous Organization: The Value of Happiness in the Workplace, *Organizational Dynamics*, 33(4), 379–392.
26. Ricardo Semler (1989), Managing without Managers, *Harvard Business Review*, September–October, p. 77.

27. See Ralph Stayer (1990), How I Learned to Let My Workers Lead, *Harvard Business Review*, pp. 1–11 (reprint 80610); see also James A. Belasco and Ralph C. Stayer (1993), *Flight of the Buffalo: Soaring to Excellence, Learning to Let Employees Lead*, New York: Warner Books; Ralph Stayer (2009), *How I Learned to Let My Workers Lead*, Harvard Business Review Classics.

28. Ralph Stayer (1990), How I Learned to Let My Workers Lead, *Harvard Business Review*, November–December, p. 10 (reprint 90610).

29. Todd A. Finkle (2011), Richard Branson and Virgin, Inc., *Journal of the International Academy of Case Studies*, 17(5), 117–118.

30. Tony Bingham and Pat Galagan (2011), M'm M'm Good: Learning and Performance at Campbells, *Training + Development*, March, pp. 36–43.

31. Pop artist Andy Warhol (1928–1987) famously predicted that in the future everyone would enjoy fifteen minutes of fame.

32. Tony Bingham and Pat Galagan (2011), M'm M'm Good: Learning and Performance at Campbells, *Training + Development*, March, p. 41.

33. For details read Art Kleiner and Vikas Sehgal (2010), The Thought Leader Interview: Vineet Nayar—The CEO of HCL Technologies Describes How He Focused His Company on Growth by Engaging Staff in Unprecedented Ways, *Strategy + Business*, 61, 114–122; Vineet Nayar (2010), *Employees First, Customers Second: Turning Conventional Management Upside Down*, Boston: Harvard Business School Publishing.

34. Art Kleiner and Vikas Sehgal (2010), The Thought Leader Interview: Vineet Nayar—The CEO of HCL Technologies Describes How He Focused His Company on Growth by Engaging Staff in Unprecedented Ways, *Strategy + Business*, 61, 114–122; Vineet Nayar (2010), *Employees First, Customers Second: Turning Conventional Management Upside Down*, Boston: Harvard Business School Publishing.

35. Ricardo Semler (1993), *Maverick! The Success Story Behind the World's Most Unusual Workplace*, New York: Warner Books. See also Lawrence M. Fisher (2005), Ricardo Semler Won't Take Control, *Strategy + Business*, Winter, p. 41.

36. Brad Wieners (2004), Ricardo Semler: Set Them Free, *CEO Insight*, April 1.

37. Ricardo Semler (2007), Out of This World: Doing Things the Semco Way, *Global Business and Organizational Excellence*, July–August, pp. 13–21.

38. Richard Semler (1989), Managing without Managers, *Harvard Business Review*, September–October, pp. 76–84.

39. Ricardo Semler (2007), Out of This World: Doing Things the Semco Way, *Global Business and Organizational Excellence*, July–August, pp. 13–21.

40. Ricardo Semler (2007), Out of This World: Doing Things the Semco Way, *Global Business and Organizational Excellence*, July–August, pp. 13–21.

41. See the various discussions in Benjamin Schneider, Mark G. Ehrhart, David M. Mayer, Jessica L. Saltz, and Kathryn Niles-Jolly (2005), Understanding Organization-Customer Links in Service Settings, *Academy of Management Journal*, 48(6), 1017–1032.

42. Also see Rhian Silvestro (2002), Dispelling the Modern Myth: Employee Satisfaction and Loyalty Drive Service Profitability, *International Journal of Operations and Production Management*, 22(1), 30–49.

8

Service in and by
Public Sector Organizations

Service in and by the Public Sector

Traditionally, public service was to the crown or to the nation-state. In this model there is no public service element as such; rather, the service is provided to the authority figures within the state and to the areas under their control. In recent years public service has undergone a shift in emphasis, with a stronger focus on efficiency of processes leading to mechanisms to deliver good service. In the early 1990s the book *Reinventing Government* described and exemplified a framework designated new public management (NPM).[1] Authors David Osborne and Ted Gaebler reconsidered structures and processes employed in private sector organizations and formulated these with a view to transforming public organizations "into a business-like identity."[2] In essence, NPM envisaged a radical refocusing of public service delivery from bureaucratic rigidity to cost-effectiveness. Although there are exceptions, conventional environments of public sector work tend to be within rule-based processes intended to standardize service throughputs and outputs. This is to help ensure that standardized public services are provided to end users who are mostly (but not exclusively) the country's citizens and taxpayers. The NPM framework envisages a separation between policy making and service delivery.[3] Since its inception, the NPM model has benefitted from a number of improvised developments, particularly at the moments of truth.[4] However, the jury is still out on whether NPM represents a real paradigm shift or is an incremental development of past practices.[5] What is noticeable is that the NPM model of public services has shifted toward a more sensitive awareness of user needs. This can be seen in the decentralization of administrative functions and market-driven approaches to procurement, supply, and delivery of some public services.[6] Inherent in public management reforms envisaged by NPM are issues of development of employees' competencies, especially of officials and bureaucrats at senior levels in governmental organizations.[7] Private sector organizations have seen similar developments stretching back over the past several decades. However, private sector environments need to meet external drivers of change, such as increased competitive pressure from globalization, the need for business continuity and survival, and the quest for sustainable profitability. Within these pressures and parameters, organizations in the private sector have comparatively greater freedom to choose what goods and services they will supply to whom and at what price. The decision of what to produce and supply is "not premised on the client's capacity or willingness to pay."[8] Different motives and perspectives influence public sector enterprises. Stringent regulations govern public sector work, and the increased focus on cost efficiency gives public service a different operational framework from private sector organizations.

Moments of Truth in Public Service Delivery

For Ernest Hemingway, the moment of truth provides satisfaction for the audience of the bullfighting spectacle. In the public sector, defining the client is less straightforward.[9] In the provision of public services, satisfaction includes taxpayers whether or not they are actual users of the proffered services; not all citizens and taxpayers use maternity services, for example. Delivery of effective service in the public sector also involves stakeholders outside the immediate service transaction. These include, for example, public policy makers, ombudsmen, and pressure and public interest groups. Public sector services are provided either universally or to specific subsections of the population.[10] Services such as taxation, policing, and judicial services are provided to the whole population. Some elements of these activities are focused on specific individuals (such as suspected lawbreakers). The aged, the sick, and the young are provided services such as state pensions, medical care, and education. Outsourcing of specific services and privatization of previously publicly owned enterprises are options. However, this is tempered by the overarching need for quality in an environment where public accountability matters.[11]

It is helpful to have a clearer idea of the composition and completion of the moment of truth—especially as the various participants are likely to have differing perceptions of the process and outcome. This issue is highly relevant for public sector work in which organizations are tasked to deliver quality service at a manageable (justifiable) cost, e.g., within set budgets. In public sector work there can be a paradox of service focus. This gives rise to a conflict between two sides of service. On the one side is the need to provide good service to consumers. This gives opportunities for moments of truth. As discussed earlier in this book, consumers contribute to good service in a symbiotic relationship with the service provider. On the obverse side is the need to deliver key functions of the public administration; obvious examples relate to community care, law enforcement, and the judiciary. And, again, key functions can also provide moments of truth. However, the target for this moment of truth might be the benefit of society at large and not solely the individuals immediately exposed to the service provision (i.e., the needy, the criminal, law enforcement officers). Here, a service such as apprehending suspected lawbreakers and subjecting them to the judicial processes is carried out in absentia (but on behalf) of the receivers of the service, the society at large. We note a direct link with Hemingway's description of the moment of truth in which satisfaction is felt by both the matador (in killing the bull) and the audience viewing the spectacle. As the object of the tragedy, the bull is presumably less satisfied with the eventual turn of events. From this perspective, the metaphor of the moment of truth is closer to Ernest Hemingway's description than the traditional way of using the concept in service management. We note also the symbiosis between matador and animal and also the role of the wider audience at the *corrida de toros* as observers and assessors of performance quality (*¡Olé!*). Delivery of public services incorporates a number of stakeholders. Stakeholders play various roles, including predicting what public services are required and what will be the acceptable levels of quality, although this can invariably be a factor of resource availability and allocation. This role is traditionally undertaken by government departments that may or may not also engage in specific service delivery. Monitoring and assessing service quality may be assigned to another government department. Dealing with complaints about quality may be the role of an ombudsman, if the civil bureaucracy has such an official within its public services structure.

In public service delivery local and national government organizations each play a role to a greater or lesser degree, and thereby provide a number of checks and balances on the delivery of public service. In terms of our current focus on moments of truth, the inclusion of government entities into the service models increases the potential for moments of truth, as there are now more entities (stakeholders) to satisfy (hence the increased number of moments of truth). For example, public service delivery is obligated to satisfy representatives of local and national elected officials *plus* the various mechanisms of state intended to ensure delivery of public services to a population. Public services make an easy target for politicians to demonstrate their own effectiveness in improving the quality of life of their constituents. There are a number of contact points for the customer (i.e., the interface between citizens and government officials). At these points the citizen (service user) has opportunities to experience public service and assess its quality. Unlike delivery of service by private sector organizations, in public service delivery the customer tends to have few, if any, choices of alternative service providers. There are, however, a number of avenues for customers to express their satisfaction or dissatisfaction with a service, such as writing or talking to members of parliament, joining a pressure group, and complaining in person to the local office. As politicians know too well, in public sector service delivery dissatisfied customers can turn into dissenting voters.

In public sector services, service processes are more complex than in the private sector. Public sector organizations differ from those in the private sector through the means by which they close the gap between the front-end personnel and the moments of truth in relation to the organization's political structure. Of course, members of parliament and other elected representatives are also themselves consumers of public services. They will thus have direct experience of the moments of truth. However, for members of the central legislature, there will be a much greater distance between themselves and the consumer. This is because there will be many organizational levels between these officials and the consumer. This gap can be narrowed by decentralizing the government structure with elected local parliaments at regional and local community levels. In turn, this will make it easier for the elected representatives to internalize the service concept.

In the private sector there is a greater emphasis on the company's role in the society. For some years now, concepts like corporate social responsibility (CSR) have been introduced as part of this perspective and have become enshrined in consumer protection legislation. Here good service is in the vanguard. Such a focus is of equal importance in the public sector as in the private sector. We should expect CSR concepts to be well distributed throughout public sector departments and their work. In fact, concepts such as CSR are arguably of more relevance given that public services are funded by taxpayers (the public purse). Automation offers many possibilities here. Against this background there is a need for good service and also for sound management practices that ensure the delivery of good service to satisfy the differing needs of the various stakeholders. To the public mind, service by public sector organizations has been typically driven by internal production capacities rather than the needs of external stakeholders such as customers, i.e., services driven by what can be supplied rather than market demand of what is needed. In most cases, public service needs will be expressly stated (such as the need to distribute welfare benefit payments). In other cases, there will be an implicit need (such as the need to develop the nation and its public service infrastructure). In the public sector, there is a difficulty in achieving consistently good service. This is especially so when components of the public bureaucracy are concerned with, for example, tax assessment and collection, law enforcement, and health. In these instances, the need for good service is driven by the needs of the various stakeholders, of which (ideally) the consumer is of paramount concern. In the public sector too, where profit generation is not a leading concern insofar as it may be proscribed, good service may be difficult to produce. Budgetary factors and resource allocation, for example, may constrain optimum public service delivery. The measurement of outputs in terms of financial cost also constrains the application of concepts from the private sector.[12] However, when high levels of quality public service can be reached, budgets become easier to achieve and more straightforward explanations can be used to satisfy concerns of taxpayers (e.g., about issues such as fairness, equity, and value for money).

Managing Moments of Truth in Public Service Delivery

There will always be a need for public sector work. And there will always be a need for managing moments of truth in the context of public service delivery. Revisiting the various attributes of these moments indicates features that need to be addressed by public service providers. The myriad moments of truth in service delivery show the complexity of preparation for consumer encounters. While each moment of truth is fundamentally unique, each has features in common. As described above, at each moment of truth a frontline employee provides service in a real-time situation—a situation wherein lie opportunities to satisfy consumers' wants and needs. For the consumer this moment affords the opportunity to evaluate service quality. The consumer's expectations (based on hearsay, recommendation, or prior experience) are a key feature of a moment of truth.

The service provider can treat each moment as an opportunity for immediate consumer feedback. Innovations in service often emanate from encounters with consumers at the moment of truth.[13] The notion that service provision is a social act (Richard Normann, 2000) indicates that training frontline staff in social skills is a minimum prerequisite for service providers. This is especially critical as the employee-customer interaction influences the consumer's perception of the organization. In this sense, and others, the customer-facing employee is the face of the organization and its ethos of service delivery. When a customer has no macro-level cues (such as brand identity), the sole representative of the organization is the person facing him or her who is delivering customer service at that time. When the customer is attracted to the service by a brand presence, the responsibility on the service-providing employee is more intense.

Training and motivating employees in the skills to manage the moment of truth is therefore a key task for executives and managers in organizations where service is a key feature of their transactions.[14] The immediate (often unique) nature of service provision emphasizes the need for quality to be "right first time, every time." This goal, aligned to the need for social skills development of relevant employees, indicates key issues for human resources (HR) development toward providing customer satisfaction through service delivery. When, as in some services, there is simultaneity of production, provision, and consumption of service, this adds to the complexity of preparing employees to deliver quality service. Adding to this complexity is an often time-critical dimension. The fact that customers invariably contribute to the co-creation of the process of service delivery affords the opportunity for immediate feedback. This is not to overlook that the underlying need is to ensure that each moment of truth provides customer satisfaction.

In public services, the delivery mechanisms are often monopolistic. This makes it difficult for the consumer to compare and evaluate quality. This is one feature of public service that is invariably more complex (and more difficult to manage) than service delivery by private sector organizations where dissatisfied customers can often migrate to competitors. One possible means to address this issue in the public sector is to create competitive systems within the public sector. Sometimes a government or industry regulator acts in ways resembling a competitor, for example, by setting regulatory requirements and benchmarks for quality standards. This is a fairly common approach when the public service relates to health and safety issues, such as the provision of clean water and sewerage services. In a more unconventional model, parallel service providers would be established within the public service network. The user/citizen thus has choices about which service to use based on publicly available data reporting quality and performance measurements. Local government agencies can allocate different benefits to the service providers according to performance and user/citizen preference (measured over time). Providers of what is judged to be good service might, for example, be rewarded by funding for training and development.

In general, there are valid rationales for shaping features of public sector organizations to be more like those in the private sector. However, as earlier discussed, the differing natures and perspectives of these two types of organizations bring different problems in service delivery. In many aspects of service delivery (whether in public or private sector organizations) there seem to be inherent difficulties in nurturing a creative climate and developing a sincere culture of service. This is especially so in the public service sector. There is a great need for creative thinking in the establishment of alternative models for public service provision. In some areas of public sector work, it is relatively straightforward to replicate service models from the private sector. In other areas, it is much more problematic. Parts of the difficulty arise from the politically charged nature of public services. It is relatively easy to privatize some public sector organizations if the nature of their work is similar to activities found in the private sector. A fundamental condition for this similarity should be that the organization is driven by similar types of incentives, such as there exists (or conceivably could exist) a natural competitive market for the service provided.

In other areas of public service, such a similarity is neither apparent nor obvious. Typical examples are the police force, the prison service, and national defense. Education, health services, and national infrastructure are also mainly in this category. There are private sector models for delivering some aspects of education, healthcare, and prison services (such as transportation and temporary holding facilities). And there are functioning private alternatives to national infrastructure and transport logistics. Using education as an example helps illustrate the complexity and managerial difficulties. In an ordinary competitive market, the well-informed and rational consumer can (at least theoretically) choose between many alternatives by evaluating price-quality differentials. And before paying for the goods or service, a customer can often try out alternative offerings, or alternatively, listen to the experiences of friends. In education, this is not possible. The time lag between service delivery/use and subsequent benefit is often measured in lengthy time periods such as months, years, or even decades. As behavioral psychologist B.F. Skinner allegedly remarked, "Education is what remains after all that has been learned has been forgotten." An additional complexity is that payment for the education usually needs to be in advance of the services provided. At the macro-economic level funding is required, for example, to build and furnish schools, train and hire teachers, and pay for similar necessary support. At the micro-economic level a school needs to assess needs of potential students before purchasing equipment such as textbooks and other resources. Given the deferred benefits of education, it is difficult to experience education services beforehand. Obviously, there is still scope for private initiatives in education. The historical record shows that many innovations in education have occurred in the private sector. Similarly, privatization in the education sector must seek to stimulate new means of education practice.

Reforming public sector infrastructure is an area where a government must take initiative and provide initial funding. This done, the work of design, construction, management, and quality assurance can be carried out by private sector organizations. This situation is not dissimilar to other parts of the public sector. Each part of the sector needs its own model. It is inappropriate to merely replicate the infrastructure from one part of the system and expect this to work effectively in other situations. From this perspective, it is especially difficult to find acceptable solutions for such public services as law enforcement, prisons, and national defense. Key issues here include the need for all members of society to benefit from the system of each of these public services, while a smaller subset of the community is directly affected by the services provided.

The Future of the Public Sector Work and the Potential of Automation

The moment of truth also has relevance in an automated version of service provision. While the majority of moments of truth continue to be face-to-face encounters, encounters facilitated by technology can bypass the personal encounter. As mentioned, usability is a key concept in the design of a technology interface to people.[15] Incorporating the moment of truth into technology design would help obviate this. In e-government, where a large proportion of the total population of a nation should be able to access the automated government services, high usability becomes even more critical. If not, the users may become disheartened with the public services. Most people know from personal experience that obtaining service from a technological provider instead of a human service provider can bring many frustrations. Whether it's the self-service auto-mat that doesn't deliver the product or doesn't deliver the expected money in change or the automated customer inquiry service at the bank that puts the caller on hold for many minutes, the technology-driven moment of truth falls short of satisfying the customer. In Chapter 4 we mentioned the 3-7-11 rule for service (if the service provider allows eleven minutes to elapse, the customer will leave and won't return).

There are many examples where public service delivery utilizes technology for self-service. In many countries, a marked success story has been the electronic provision of revenue services via the Internet to access relevant information, guidance, and documents.[16] In a developing economy such as Thailand, the e-revenue service has been lauded as a successful transition toward delivering service to citizens. Another example is in the administration of labor markets, where a key function is the placement service for job seekers. In some countries this function consumes a very large proportion of public service resources, including work time by relevant officers. In many countries, and very early in countries such as Canada and Sweden, face-to-face contact was transferred to Internet self-service.[17] In the past, when the manual service in this respect was excessively time-consuming, it was difficult for personnel to focus on more important and critical tasks, such as giving advice and support to people with special needs, such as the long-term unemployed and the handicapped. The introduction of e-service, with its high component of self-service, has freed up time for officers to focus on special needs where consumers benefit from face-to-face advice, which includes empathy. Instead of more routine types of work, personnel can be engaged in moments of truth of high complexity, and the service thereby provided fulfills important social and caring functions. This is a very good example of how automation in the delivery of service can change the characteristics of the moment of truth, and how the moment of truth can change the quality of service. This exemplifies how a new emphasis on the service provision at the moment of truth brings added social value for the consumer and society. The cost-benefit of this focus on service delivery will be enormous.

Other areas of automation are related to what is often called e-citizen, or efforts to improve and facilitate the process of democracy in the society. Local direct participation in decision making can become an everyday reality and bring the citizen closer to the processes of public service provision.[18] This means a creation of a new type of moment of truth that, until now, has been unavailable in the traditional representative parliamentary democratic system. Instead of elaborate time-consuming representative processes, the citizen can access online and be a part of real-time decision processes.[19]

Another area where technology is applied to public service is e-learning.[20] The phenomenon of e-learning is a complex application of the moment of truth. The phenomenon has two distinct levels. At the one level is the online provision of training to public sector officers and employees. Here the focus is development, training, upgrading of skills, and development of new skills. At the second level is the interface between the citizen and the public provision of education, advice, and support for citizens.

In concluding, we note that management educators are advised to internalize good service. Management educators whose work incorporates educating business professionals need to learn from the teachings of their own classrooms. In order to fulfill their role as management educators, they need "to develop a customer orientation and must also become expert generalists."[21]

Endnotes

1. David Osborne and Ted Gaebler (1992), *Reinventing Government: How the Entrepreneurial Spirit Is Transforming the Public Sector*, Reading, MA: Addison-Wesley.
2. Per Skålén (2004), New Public Management Reform and the Construction of Organizational Identities, *International Journal of Public Sector Management*, 17(2–3), 251–263.
3. Jennifer Rowley (1998), Quality Measurement in the Public Sector: Some Perspectives from the Literature, *Total Quality Management*, 9(2–3), 321–333.
4. Christopher Hood and Guy Peters (2004), The Middle Age of New Public Management: Into the Age of Paradox? *Journal of Public Administration Research and Theory*, 14(3), 267–282.
5. See, for example, assessments and discussions in Laurence Lynn (1998), A Critical Analysis of the New Public Management, *International Public Management Journal*, 1(1), 107–123; Laurence Lynn (2001), The Myth of the Bureaucratic Paradigm: What Traditional Public Administration Really Stood For, *Public Administration Review*, 61(2), 144–160; Christopher Hood and Guy Peters (2004), The Middle Age of New Public Management: Into the Age of Paradox? *Journal of Public Administration Research and Theory*, 14(3), 267–282; Stephen Page (2005), What's New about the New Public Management? Administrative Change in the Human Services, *Public Administration Review*, 65(6), 713–727.
6. Stephen Page (2005), What's New about the New Public Management? Administrative Change in the Human Services, *Public Administration Review*, 65(6), 713–727.
7. Christopher Hood and Martin Lodge (2004), Competency, Bureaucracy, and Public Management Reform: A Comparative Analysis, *Governance*, 17(3), 313–333.
8. J. Teicher, O. Hughes, and N. Dow (2002), E-Government: A New Route to Public Sector Quality, *Managing Service Quality*, 12(2), 391.
9. John Alford (2002), Defining the Client in the Public Sector: A Social-Exchange Perspective, *Public Administration Review*, 62(3), 337–346.

10. J. Teicher, O. Hughes, and N. Dow (2002), E-Government: A New Route to Public Sector Quality, *Managing Service Quality*, 12(2), 391.

11. See L. O'Toole and K.J. Meier (2004), Parkinson's Law and the New Public Management? Contracting Determinants and Service-Quality Consequences in Public Education, *Public Administration Review*, 64(3), 342–352.

12. See J. Teicher, O. Hughes, and N. Dow (2002), E-Government: A New Route to Public Sector Quality, *Managing Service Quality*, 12(2), 384–393.

13. See, for example, Marc Beaujean, Jonathon Davidson, and Stacey Madge (2006), The Moment of Truth in Customer Service, *McKinsey Quarterly*, 1, 62–73.

14. G.R. Bitran and J. Hoech (1990), The Humanization of Service: Respect at the Moments of Truth, *Sloan Management Review*, 31(2), 89–96.

15. Brian Hunt, Patrick Burvall, and Toni Ivergård (2004), Interactive Media for Learning (IML): Assuring Usability in Terms of a Learning Context, *Education + Training*, 46(6/7), 361–369.

16. See Stuart J. Barnes and Richard Vidgen (2004), Interactive E-Government: Evaluating the Web Site of the UK Inland Revenue, *Journal of Electronic Commerce in Organizations*, 2(1), 42–63; Regina Connolly, Frank Bannister, and Aideen Kearney (2010), Government Website Service Quality: A Study of the Irish Review On-Line Service, *European Journal of Information Systems*, 19(6), 649–667.

17. See discussions in Richard T. Cober, Douglas J. Brown, Alana J. Blumental, Dennis Doverspike, and Paul Levy (2000), The Quest for the Qualified Job Surfer: It's Time the Public Sector Catches the Wave, *Public Personnel Management*, 29(4), 479–496; Karen E. Pettigrew, Joan C. Durrance, and Kenton T. Unruh (2002), Facilitating Community Information Seeking Using the Internet: Findings from Three Public Library–Community Network Systems, *Journal of the American Society for Information, Science and Technology*, 53(11), 894–903. Also see relevant sections Arvo Kuddo (2012), *Social Protection and Labor: Public Employment Services and Activation Policies*, World Bank Discussion Paper 1215.

18. Toni Ivergård and Brian Hunt (2005), Towards a Learning Net-Worked Organization: Human Capital, Compatibility and Usability in E-Learning Systems, *Applied Ergonomics*, 36, 157–164.

19. Karl W. Sandberg, Toni Ivergård, and Stig Vinberg (2004), E-Service to Citizens and Companies in Rural Areas, *International Journal of the Computer, the Internet and Management*, 12(2), 213–223.

20. Toni Ivergård and Brian Hunt (2005), Towards a Learning Net-Worked Organization: Human Capital, Compatibility and Usability in E-Learning Systems, *Applied Ergonomics*, 36, 157–164.

21. K.J. Blois (1992), Carlzon's *Moments of Truth*: A Critical Appraisal, *International Journal of Service Industry Management*, 3(3), 16.

9

Public Sector Culture and Values: Delivering Public Service Excellence

In this chapter we describe a government agency whose role is to administer the national labor market. This case study is based on the personal experience of one of the current authors (TI). We describe and illustrate the agency's structure, work processes, and personnel. In combination, these attributes shape the agency's organizational culture, which in turn influences the productivity and performance of its exceptionally dedicated employees. Of particular relevance are issues of performance measurement, especially how the agency chooses to view and measure success. At the core of the agency's work is its effectiveness in placing job seekers in employment. In essence, this is a key performance indicator (KPI) for the agency as a whole and for its officers. Success can thus in part be measured through customer satisfaction: Is the employer happy with the person who fills the job vacancy, and is the job seeker happy in the job he or she receives, or does the agency receive frequent requests to arrange changes of jobs? We would term this the macro-level of the agency's work. At the micro-level is the work effort and performance of individual officers in working with their job-seeking clients, for example, to provide work-related counseling and to place people in suitable seminars and workshops.

The Work and Structure of the Agency

Located in a country in Northern Europe, the agency is responsible for administering the national labor market. Within this overarching brief, the agency manages the fluctuations in labor demand and supply. On the demand side, the agency works closely with local employers to identify and advertise job vacancies. On the supply side, the agency works closely with job seekers and the long-term unemployed. In this part of the work, the agency designs and implements recruitment processes to match job seekers with advertised positions. In some cases, job seekers who are otherwise qualified "on paper" may lack the personal skills necessary to attract employers. This may be prevalent in the long-term unemployed whose lengthy period out of work has lowered their self-esteem and social adeptness. In such cases the agency conducts appropriate training, for example, in confidence building and social interaction for successfully sitting and passing a job interview. It may also be necessary to provide briefings and workshop sessions on completing job applications, preparing curriculum vitae (CV), basic computer literacy, timekeeping, and issues connected with health and safety. In these ways the agency plays a proactive role in preparing job seekers for employment. The agency's brief also includes administering processes for the provision of benefits to the unemployed.

The agency is organized in a nationwide network of employment offices. In effect, these offices form a customer-facing service function with clients. In the local offices, the agency's officers are recognized for their expertise and are acknowledged representatives of the central governmental authority. As such, the officers may be called up to explain policies and changes in labor regulations. As would be expected, the agency operates a disproportionately high number of offices in areas of high unemployment and, where possible, staffs these offices with highly competent and experienced officers. In this way the agency maintains a large operational presence in areas where reside most of its clients (people who require employment-related products and services) or large numbers of customers with special social service needs (such as the disabled or physically impaired). In areas where the labor market is more stable, the agency tends to have a smaller presence: fewer offices or smaller offices (manned by fewer officers). For job seekers, the agency has a range of services: identifying and publicizing job vacancies, identifying unemployed people as potential applicants, interviewing suitable candidates for the vacancies, providing counseling, and conducting suitable training. The task of "matchmaking" (linking applicants to vacancies) is mainly done via databases of people "on the books" and "walk-ins" (people who visit the agency requesting job-related help). The next step of preliminary interviewing has traditionally been by personal face-to-face interaction, but is nowadays through electronic means of communication. Several stages of the processes can now be conducted via electronic means.

The national network comprises approximately 340 local employment offices. All of the major cities have several employment offices. Most of the nation's municipal centers also have at least one employment office. In terms of its overall network structure, the agency is decentralized: more officers work in the regions than in the capital city. The vast majority of operational decisions are made by the agency's officers in each of the regional and local offices. There is little need to refer to centralized administrators for operational decisions. Local decisions tend to address issues of day-to-day importance to job seekers, such as transport to the workplace, financial assistance before receiving the first paycheck, and family-related matters. In the main, such decisions are tactical: needing immediate resolution within a relatively short-term timeframe. This type of operational decision is often balanced by decisions that have a strategic dimension. These too are made locally, for example, planning activities to alleviate unemployment, proactive initiatives with local employers to predict job vacancies and thus preempt seasonal unemployment, and longer-term decisions intended to create jobs throughout the region. Like their colleagues in other areas of social work, the agency's officers work with clients on a case-by-case basis, a feature that gives a personal dimension to the agency's work processes. In the process of helping clients, the first stage is to identify the client's employment situation and immediate needs. Here, high levels of empathy and sensitivity are needed to match the available vacancies with the client's stated needs or wants. The majority of the work of the officers could be termed emotional work.[1] For the individual case officer success comes from matching client needs to wants. This is also a key source of satisfaction for both the officer and the client. In general, an officer's experience, expertise, and training guide the portfolio of cases that he or she deals with. Some client cases are routine, while others are more complex and more challenging for the officer or officers involved.

From around the late 1990s up to the current time, the agency has maintained a relatively stable full-time workforce of around 12,000 officers. On average, each officer has a minimum of ten years public service. However, the vast majority of officers have long records of professional service either with the agency or in a similar type of public service. It is not unusual for some officers to serve their entire working life in the agency. The agency tends not to be a workplace with high rates of turnover. Employees in public service organizations are characterized by high levels of personal altruism and commitment to serving the public good. Such employees have high levels of job satisfaction and low levels of a desire to leave their organization when there is a close fit between their own personal values and their organization's values.[2] By organizational design, the agency is highly decentralized. By far the greater proportion of full-time officers serve in the agency's regional offices. In terms of its organizational structure, the agency is very flat. Overall, there are four main function levels from the director general (DG) (based at central headquarters in the capital city) to the frontline (customer-facing) officers. The levels are the DG, labor directors, office directors/managers, and frontline desk officers. In the regions, the agency's offices normally have three or fewer levels of administration. One of the agency's underlying strengths is its flat structure and a concerted avoidance of a hierarchy. The organizational feature helps the officers deliver fast and effective service to meet their clients' needs.

The agency's director general (DG) is based in the capital city and reports directly to the Minister of Labor. The DG has administrative and political responsibility for the whole agency. At the next level of responsibility are the labor directors. These officers are based in the regional office and oversee the work of the local offices in their regional area. Depending on their size and location, each agency office is under the stewardship of a director or office manager. Larger offices with a director are located in areas of high unemployment where the office needs to provide a wider range of services

FIGURE 9.1
The agency in relation to its clients and headquarters.

to clients. Figure 9.1 shows the agency in relation to its clients (local citizens), the headquarters, and the labor market.

A key function of the central government and its ministries is to provide overall policies, targets, and budgets as appropriate to the perceived labor market needs in a given area of the country. Naturally, these are based on political realities and the pragmatism of government. In terms of its working mandate, the agency is, in itself, self-sufficient. It is tasked to address issues in the labor market and to formulate actions that address these issues. That said, the central government exerts continuous pressure on the agency to improve productivity and efficiency. Key performance requirements are fourfold: active labor market goals, better service to the nation's citizens, reduce the numbers of agency personnel, and reduce budgetary expenditures. Each of these four areas allows the government to demonstrate its awareness of societal needs and to state that its policies are on track. Two of the four performance requirements (reduce personnel numbers and reduce expenditures) comply with the instructions of governments worldwide to their ministries to obtain more output from fewer resources. Almost year on year, the agency succeeds in fulfilling each of these set targets. This makes it somewhat unusual. In many countries, many government agencies fail to meet these target areas.

A client's first encounter with the agency is with a frontline officer. For the most part, these officers are trained and experienced in resolving clients' unemployment issues. More junior or newly recruited officers may be learning on the job with the help of a mentor. The agency's flat organizational structure means that more senior or more experienced officers at the supervisory level can make operational decisions in close coordination with their colleagues who work directly with clients. In each agency office colleagues can therefore discuss a client's problems in real time while the client is nearby, which helps ensure fast resolution of problems. It also helps ensure that clients do not leave the office disappointed with the services they have received.

On occasions, supervisors may need to consult with their regional directors. In workplaces with relatively few employees, the close proximity of supervisory staff aids swift and open communication, effective decision making, and a friendly environment for the public service work. A combination of a flat organizational structure, competent and empowered officers, and a prevailing ethos of trust and mutual assistance between workplace colleagues contributes to effectiveness (appropriate and speedy resolution of clients' problems) and efficiency (cost-effective use of resources). A flat organization and experts in close proximity also help make communication straightforward and prevent distortion and misunderstandings. In a workplace where employees work closely together, word-of-mouth communication keeps everyone on the same page and helps ensure that everyone is well informed. The short distance (in both time and space) between problem identification (words) and problem resolution (actions) aids service delivery. Thus, there is little scope for error, wasted time, or duplication of effort.

As mentioned, many officers devote their whole working life to the agency. At the agency expertise and experience are highly prized. Veteran officers are appointed to key administrative posts where their knowledge and accumulated skills can be utilized in educating and mentoring junior colleagues. On joining the agency new recruits are assigned to a senior mentor for several weeks before joining their first posting to a regional office. Organizational mentoring systems tend to nurture trust and empathy between work colleagues as well as providing a long-term framework for problem solving.[3] Mentoring systems also supplement induction programs that introduce new employees to the processes, systems, and culture of their employer. In the early days of the agency a majority of its officers joined from working in trade unions (and predominantly from blue-collar unions). Over time, this traditional route into the agency has given way to a more systematic hiring process that takes account of professional training and competencies and often academic performance. Nowadays, most officers hold university degrees in, for example, human resource management, business, or law.

Delivering Public Service: A Political Dimension

A key strand that runs through public service provision is the political dimension.[4] And the public services provided by the agency are no exception. The agency's headquarters are located in the government district in the capital city. The area contains a number of ministries; the Ministry of Labor is close by, as is the office of the prime minister. The agency does not have a "fat" administrative division; its approximately 400 administrators represent about 3 percent of its total workforce. The work of the headquarters' personnel invariably focuses on national issues (the big picture). Important tasks include setting employment policies and coordination of effort with the work of other ministries. Setting policy in lockstep with the initiatives of other government departments and ensuring compliance with political agendas influence the work of the officers at the agency's headquarters. Accurate reporting of employment data and being "on message" with political statements are key success factors for these officers. Against this backdrop, the agency is designed to be independent from the formal political arena. This means the agency's officers advise rather than set policy, and inform rather than direct the work of ministers. By its founding charter the agency is intended to take a neutral stance and focus on delivering excellent public service with the aim of addressing labor market needs and rectifying imbalances in national labor demand and supply. Not surprisingly, headquarters-based officers need to balance empathy with *realpolitik* and to juggle the demands of ministers (career politicians elected for a term of office on the basis of a political manifesto), the regulatory frameworks of the agency and the civil service, and the evident social needs of clients. It is no easy task to juggle each of these contrasting demands. And with their function to oversee a widespread network of agency offices, headquarters officers are closely involved in designing instruments for performance measurement. This too can have a political dimension. Notoriously, politicians are ever vigilant for data and information that show their value to the electorate and the success of their policies.

The Agency's Organizational Culture and Values

By definition, corporate culture is "the way we do things around here."[5] Among other influences, such as the personality and preferences of an organization's founders, corporate culture tends to be influenced by the prevailing dominant group and its actions. Culture pervades organizations, whatever their size, shape, or disposition. The larger an organization, the greater will tend to be the variation of culture(s) among the internal subgroups.[6] Culture is said to be "more than the sum of its parts."[7] The culture of an organization has the capacity to influence behaviors and activities within that organization. Where employees share the same background and thus cherish the same values, these influences can be positive.[8] Influences can be negative, as in organizations where employees are punished if they do not conform to prevailing norms.[9]

Organization culture plays a role unifying potentially diverse behaviors of employees. Where workplace behaviors are predominantly shared by employees, the organizational culture is said to be strong.[10] Organizations in which the employees do not share similar views toward their organization's values are said to have a weak culture.[11] In workplace environments where employees' values and the organizational values are a close fit this is described as "value congruence."[12]

The Agency as a Cadre Organization

The political scientist Bo Rothstein classified the agency as a "cadre organization."[13]

The distinctive features of a cadre organization are often found in political, religious, charitable, or volunteer organizations: a sense of mission, a belief in the worth of the mission, and a willingness to pull together with colleagues to implement the tenets of the mission. A key attribute of cadre organizations is their ability to be highly responsive to making organizational change. In the view of Rothstein, the agency had a very high level of responsiveness to change, whether this was in its ability to implement new ministerial policy directives or to perceive changes in the labor market, such as the need for new skills and competencies by employing firms and people seeking jobs. At the macro-level of the national labor market this means that the agency can quickly implement stated policies (decided at the political center) into actions (carried out locally). With this level of innate fast responsiveness embedded in the agency's organizational structure, it was invariably left to the agency's executives and managers to design and direct any required process of change. This is somewhat unusual in public sector environments (although there can be exceptions). In general, a public sector organization tends to be averse to making rapid changes either to external environments or to political pronouncements or directives. Size, structure, and especially culture are often cited as reasons for maintaining a status quo.[14] In a cadre organization "goal fulfillment" fulfills a role of intrinsic motivation for the employees.[15]

Key organizational features of cadre organizations are rarely evident in other types of organizations.[16] According to Rothstein, a public sector cadre organization most closely resembles companies in the private sector.[17] A cadre organization's members often have similar work life backgrounds and experiences, and these shared personal histories nurture a collective sense of togetherness and a mutual sense of belonging. From this shared background individual members develop empathy for colleagues and a collective sense of purpose that is well defined and often does not need to be expressed. This contrasts strongly with some organizations that post public notices displaying the organization's vision, mission, and values in a vain effort to remind employees of their roles, purpose, and goals. In some organizations lethargy, politicking, and employees' reluctance to make decisions slow down response times and prevent swift actions. Without these impediments members of a cadre organization can remain focused on their organization's direction and goals secure in the knowledge that these are shared by colleagues. An environment where employees feel empowered and know that they have the support of their colleagues builds up an *esprit de corps* while reducing interpersonal conflict and professional discord. Employee collaboration toward agreed goals allows for rapid decision making at the point where speed of decisions matters: at the interface with customers and clients.

As everyone in the agency shares a common background regardless of their level of seniority or rank, it is relatively straightforward to reach consensus on policies, goals, and actions. This too is a feature found almost exclusively in a cadre organization. In many, if not most, organizations there is likely to coexist environments that encourage employees to engage in competition or cooperation.[18] In workplace environments that value cooperation, employees share ideas, tolerate error, and engage in mutually supportive activities such as sharing resources, developing joint initiatives, and engaging in activities for mutual learning. In such organizations the prevailing workplace ethos encourages employees to seek win-win outcomes with colleagues. In work environments where competition is the norm, employees are opportunistic and workplace behaviors are a zero-sum game as employees make power plays to access resources and gain the attention, support, and preference of senior managers and executives. In such organizations employees behave more out of self-interest than in the interest of their organization or their colleagues. The prevailing culture of an organization (the ideological mind-set of the employees) shapes the behavior of employees toward mutual cooperation or competition. An organizational culture that encourages employees' cooperation will likely have a *positive* influence on employees' perceptions of their work (e.g., create high levels of job satisfaction). An organizational culture that encourages employees' competition will likely have a *negative* influence on employees' perceptions of their work (e.g., create a high rate of employee turnover).

In public sector cadre organizations employees have high levels of public service ethos and regard duty and their provision of public services to citizens as paramount motivations for their work efforts. Altruistic motives take precedence over self-centered driving forces of human nature, such as personal self-interest, political maneuvering, and coalition building. Although possibly separated by the physical distance between headquarters and the regional outstations, the similar professional backgrounds of the agency's officers encourage a mutual understanding throughout the national network. So while there is a great distance between the agency's headquarters in the affluent capital city and the regional offices, the mind-sets and worldview of the agency's officers are similar. This leads to a ready understanding of the purpose of the work and the means of conducting day-to-day tasks. Whereas in earlier times agency employees had a shared background of union membership, nowadays they share a strong belief in public service working to provide services to people whose personal circumstances indicate that they need help and support.

The common backgrounds of members of a cadre organization help generate a shared understanding of workplace focus and possible actions. When such a high level of mutual understanding exists in an organization, it is likely that this will have positive knock-on effects. A shared understanding between employees is likely to lead naturally to feelings of mutual trust and collective effort toward commonly held goals. In turn, these features become key drivers of employee effort common outputs. In essence, in terms of workplace focus and effort, the whole organization is greater than the sum of its individual employees. To facilitate empowered decision making, the organization's operational rules tend to be guidelines for employees' interpretation rather than strictly enforceable must-do diktats. A climate of collective empathy and collaborative responsibility allows for role clarity and speedy decisions. In organizations, a shared ideology promotes cooperation, collegiality, and consensus.[19]

In a cadre organization the employees themselves drive their organization forward, and in a number of ways. First, there is a shared sense of purpose with output goals that all employees share. As, by definition, employees in a cadre organization have experienced common personal and work histories, the purpose and goals of the organization tend to be widely understood and are internalized within each employee. A second driver in the organization is the high level of mutual trust between colleagues that gives rise to a supportive work environment and contributes to an *esprit de corps*. From this *esprit* stems a sense of mission that in public service work contributes to what is traditionally called the public service ethic.[20] Indeed, work in the public sector has been called a "special calling" seeming to attract people of a particular type.[21] Public service motivation (PSM) is said to contain features not usually found in employees in private sector organizations.[22] Key attributes of public service motivation are a concern with the needs of citizens (especially citizens who are less fortunate than the public officers themselves) and a desire on the part of officers to seek ways to improve service delivery. A sense of pride in the work underpins public service motivation.[23]

The Agency's Service Delivery and Performance Measurement

From the late 1980s the agency has moved toward being goal-oriented and vision and mission driven in terms of its work outputs. In parallel, there has evolved a range of sophisticated performance measurement criteria. Performance measurement takes a two-pronged approach. At various times, the agency has used a large number of comprehensive sets of different performance measures. Some instruments are designed to measure the overall performance of the agency in satisfying labor market needs. This reflects the performance of the agency nationwide and the combined work of all officers throughout the agency network. At a more micro-level the agency gathers and measures job performance data from individual officers, and especially those in supervisory roles. Thus, a number of performance measures focus on individual performance in what is conventionally known as performance appraisal. The key means of measuring the performance of individual officers is an annual performance appraisal conducted as part of ongoing staff development routines. The results of these performance appraisals help guide and shape an employee's career, his or her work tasks, and job location.

The agency seems to recognize that there is potential and actual overlap in work performance between organizational outputs (to which employees contribute collectively) and outputs by individual employees (to which employees make an individual contribution). However, it is often difficult to disambiguate these outputs and allot them to separate sources of effort. This is invariably in the nature of performance measurement.[24] The process begins by measuring performance outcomes in each office. At this level the focus is on filling job vacancies within the local labor market (for example, how many job vacancies were advertised and how many of these were filled within a certain timeframe). Recognizing that such criteria focused on quantity, the agency's performance appraisal routines encompass individual employees and predominantly focus more on quality of service (and include feedback satisfaction data from clients and employers). Predictably, when organizations focus on measuring and assessing quality of service and find ways to reward this, then the quality of service provision tends to increase.[25]

In the local branch offices, performance is largely related to each office's effectiveness in satisfying the needs of the labor market. However, the nature of a cadre organization is that employees are closely involved in the overall performance of their organization. Employee performance is often intertwined with the performance of their branch office. In cadre organizations employees recognize that, in many ways, their individual performance *is* their organization's performance. This is not to say that individual performance cannot be appraised separately from organizational performance (for example, as part of an annual exercise of individual goal setting and personal development), but that individual employees feel a close affinity with the performance of their organization.

Recognizing the need for processes to encourage self-adaptation of individual officers to local circumstances, the agency designed a bottom-up appraisal process for most of its evaluations. In this, several important issues are to be addressed (as described below). To create this form of self-adaptation, it is important for the organization to have access to a bottom-up appraisal process. The process of bottom-up appraisal is intended to assess the contribution of individual officers' feelings to the whole organization. The agency recognized early that officers' views of their workplace climate are an important set of data. The agency has thus tried to develop sensitive instruments and methods to encourage employee feedback on the organization and the internal environment in which they work. These new methods have become important decentralized complements to existing and centralized (top-down) methods to measure efficiency and productivity of frontline activity as well as other activities of regional outposts.

Performance Measurement of the Local Branch Offices

In a labor market environment, inputs for performance measurement can come from various sources. The agency has designed processes of assessing organizational-level performance for which the primary data sources are the clients who receive the agency's services and the agency's officers themselves. For the agency's local branch offices, success means placing job seekers in suitable jobs. In this, priority is given to the unemployed (rather than, say, people in work who wish to change their job). In macro-level terms this means that individual branch offices aim to satisfy the needs of their local labor market. Data to demonstrate successful completion of this mission can come from many different kinds of indicators: for example, the speed of placement into suitable jobs of the currently unemployed, or the development of workplace competencies and skills through training programs to alleviate bottlenecks attributable to keeping people out of work. Common for all data indicators and measures is a requirement that these should demonstrate the proactive role of the agency's officers in the labor market, i.e., managing employment supply and client demands for work. However, due to regional differences, e.g., seasonality of employment or a surfeit or lack of relevant skills, there may be marked differences between the national labor market and the local labor market. At the local level the agency may need to take a more interactive (some might say intrusive) role in the local labor market, for example, by working closely with employers and potential employers to shape the profiles of job requirements.

For some time, the agency has had a process of customer feedback gained through a separate client evaluation. This evaluation is instigated and implemented by central headquarters administrators, although the local branches are responsible for taking responsive follow-up actions as appropriate. The regional branches are expected to take direct responsibility and are authorized to respond in their own ways to the survey findings. Thus, the local branches are responsible for improving their customer service where the survey identifies shortfalls in the current levels of service. Officers in the local branches develop action plans and implement these and, by routine processes, report back to the headquarters. Officers at headquarters monitor and measure outcomes and incorporate these into a national picture for reporting to the relevant politicians.

For some time systems for appraising individual employee performance have been among the agency's established work practice. Staff appraisal exercises are carried out on an annual basis by immediate line managers. Appraisals tend to take the form of a dialogue between managers and their subordinates and focus on work done by the individual staff member. One outcome of local-level appraisals is the individual work contracts that set out parameters for work, training, and rewards and remuneration payments. The work performance of local office managers is appraised by their immediate manager, who has responsibility for a number of agency branches in a given area. At this level, the focus is on performance of the branch officers in meeting stipulated targets for satisfying labor market demands (as represented by the number of job seekers and the number of people who are placed in work).

Employee Evaluation of Work Conditions (360° Feedback)

In addition to appraisal of individual employees, the agency has recognized the need for bottom-up (360°) evaluation of work conditions in the individual offices. These data are gathered through annual work climate surveys. The survey content focuses on leadership, organizational encouragement, social support, rehabilitation, feedback and evaluation, and competence development. Conducted annually by the central (headquarters) administrative staff, the survey findings are publicized nationally. Members meet together with their manager (or managers) in very small groups to discuss the survey findings and exchange ideas about the implications for work routines and practices. The members of a local work group discuss and interpret their own results and develop a plan of action. Implementation of each local action plan is the collective responsibility of the members of each work group. Local work groups thus own the action plans designed by their own work group together with the outcomes. In this way actionable outcomes are devolved to frontline officers. Table 9.1 summarizes key features of the agency's appraisal processes.

It is essential that new bottom-up processes harmonize with the existing productivity assessments of overall office performance of meeting the needs of the labor market. Traditionally, the productivity of each employment office was the basis for successful job placements. These data are seen as important public documentation within the branch offices. Productivity results are reported monthly and annually. Details are posted in the public workplace domain and are thus routinely available to all employees. The nature of the officers means that productivity figures become targets for work process improvement. In practice, officers eagerly anticipate the latest publicized figures as a way of benchmarking their own office performance against the performance of

TABLE 9.1

Performance Appraisal Processes (Individual Employees)

Performance Appraisal	Frequency and Format	Focus	Aims and Rationale
Branch offices	Annual Customer evaluation survey conducted by headquarters	Evaluate levels of service delivery from the perspective of service users	Provide a benchmark for improvements in the service; identify shortfalls in service delivery; set targets for outputs
Individual officers	Annual Dialogue between the line manager and a subordinate	Set out work parameters, training opportunities, payment systems	Match individual work performances to branch office outputs; set targets for outputs
360° feedback process	Annual A workplace climate survey; conducted by headquarters; discussion groups held with officers in branch offices	Officers self-critique their work processes and outputs, share ideas for improvement, and design an action plan for the following year	Officers in the branch offices own their work outputs and the path toward delivering these outputs; officers have intimate knowledge and involvement of their processes of service delivery

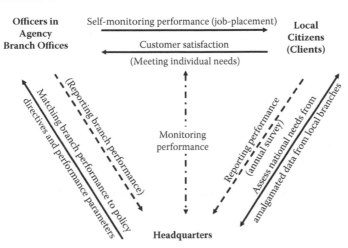

FIGURE 9.2
Inputs for measuring performance (organizational level).

colleagues in other offices in a form of league table. Publicizing performance data in this way encourages positive approaches to reflecting on workplace behaviors with a view to making improvements. Figure 9.2 shows the various data inputs for measuring performance at the level of the organization.

The agency's central senior executives, the directors general (DGs) and deputy directors general (DDGs), routinely carry out performance evaluations of each regional administration. Conducted either annually or every second year, and comprehensive and far reaching, these top-down evaluations normally include most types of measures and indicators. As a result of the evaluation, the DG and the relevant regional director sign a joint declaration whereby the regional officers commit themselves to an action plan for their province. In turn, the central part of the agency commits itself to provide resources as well as other types of support (experts, evaluations, etc.).

However, in practice, supplementary resources are rarely provided. Budgets are distributed from headquarters using a kind of needs model that allots fair distribution of resources over the whole nation. If a particular province has a special need, the first priority is to reallocate elements of the existing budget. While this may present difficulties, these can be resolved by a creative labor director. Allocating human resources to meet demand peaks is more problematic. Even so, there are degrees of flexibility as the agency's work occupies a very large part of labor market forces. Primarily, the work group's own budget is used to fund implementation. If needed, extra budget can be requested. In some parts of the organization the responsibility for this budget is held by joint worker-management work life committees. The agency has continuously developed and redeveloped elaborate models to ensure a fair distribution of financial resources to the agency's various component parts. Managers at higher levels (including local-level managers) cannot participate in the results without invitation from the local work groups. Local work groups were recommended to discuss all the different performance measures at the same time and to develop comprehensive action plans, including labor market efficiency and quality of working conditions.

Regional leaders (e.g., a manager of a local employment office or a director of a province) are expected to comply with the policies and goals set by the central (headquarters) authorities. This means that they are obliged to adjust their own working to the stated projected levels of productivity and to take into account the publicized measurement of productivity. Work adjustment by organizational leaders and individual frontline (customer-facing) officers is often in lockstep. Leaders adjust overall work priorities based on performance feedback from the labor marketplace. Individual officers adjust their own workplace behaviors on the basis of real-time feedback from their daily interactions with their clients. In a cadre organization, the closeness of professional relations (including awareness of feedback and decision making) enables groups of colleagues to adjust their collective behaviors so that their own behaviors are not out of step with those of their colleagues. Officers routinely use feedback from their day-to-day caseloads. They also adopt their work practices to announced central goals and policies. Their own perceived realities from their day-to-day work thus act as a moderating influence. Thus, their learning (broadly defined as a perceived change in workplace behavior) reflects a double-looping process.[26] As an example, a noticeable development in how job seekers preferred to identify suitable job vacancies and work opportunities was driven by their technology competences. Local offices installed computer terminals and Internet access as a service for the unemployed. This service has become remarkably popular with both the agency's customers and its personnel. The process is a form of self-service/self-help model of assisting those seeking work. Encouraging those job seekers whose work inquiries are relatively straightforward to take responsibility for finding more details for themselves allows officers to spend comparatively more time with job seekers whose needs are more intricate, complex, and difficult to resolve. The benefits are threefold: job seekers gain increased confidence in finding employment opportunities (a useful skill for the long-term unemployed), officers have more time to deal with more challenging cases of unemployment, and at the level of the branch office performance, there is potential for positive perception from clients of the ability and speed with which the local officers satisfy their job needs. In terms of the overall form and function of labor market management, this process can refocus expertise onto technological support and help shorten lead times for appropriate help for the unemployed. Over time, this may reduce the total number of agency personnel as well as making individual case work more interesting. It could lead to the setting up of a national network of job seeker booths where the unemployed could use resources such as Internet websites and a telephone hotline for employment advice. The organization's leaders have to consider the findings of the 360° feedback, that is, the feedback from agency employees on the performance of their own managers as leaders in their workplace. Sometimes, this information might be incongruous. For example, an individual office might perform well in terms of meeting labor market needs, but the opinion of individual officers might suggest that key leadership qualities are lacking.

As discussed earlier, in a cadre organization, members at the various levels in the organization exhibit high levels of mutual trust and support and have a shared sense of mission. This shared sense of mission, together with a high level of commitment, makes it possible for officers to have a deep understanding of the essence of the intention of legislation (i.e., the *spirit* of the law). Conceivably, this understanding may be more thorough than the textual content expressed in an actual legal document (i.e., the *letter* of the law). In organizations where members share ideological beliefs, these beliefs are likely to supersede formalized rules. This seems to be a paradox of a cadre organization. Over the past decade, the trend toward increased decentralization and the continued empowerment of personnel might result in an increased number of possible conflicts of interests and thus give rise to potential legal problems.

Taken together, these features tend to help ensure a commonness of goals and direction. Ideally, performance measurement should be closely linked with program goals.[27] In performance measurement it is essential to avoid any drift away from stated goals. Attention of both management and officers should be in alignment.[28] This means alignment of policy setters and policy implementers. Divergence between performance measurement and organizational goals can lead to dysfunctional behavior.[29] Goal divergence may stem from the multiple principles and multiple work tasks of public sector organizations.[30] When designing public sector policy, it is therefore important to take into account the multiple work tasks and the various principals.[31]

Public Service Excellence

The most important factor seems to be related to leadership. This includes the perceptions of officers in the quality of the leaders. Within this there are several dimensions. At the level of the local offices, the confidence that officers have in their immediate supervisor is of prime importance. In this cadre type of organization, officers are satisfied if they can feel confidence with their immediate supervisor. In some organizational forms (e.g., matrix types of organization) leadership is less than obvious; often it is opaque. Another dimension is that the supervisor should contribute to an acceptable allocation of work tasks between peers. Officers expect their supervisor to allocate work tasks fairly; if not, they will be dissatisfied. Apparently, it is also important that when deficiencies have been identified and prioritized, officers (peers) jointly contribute to improve their own working conditions. The agency's leaders are intended to have an important role in facilitating this process. When they identify inefficient procedures, officers strongly expect their manager to be proactive in making appropriate changes. This bottom-up process informing top-down action is a feature where the collective participation of work colleagues to workplace conditions is obviously noticeable.

Feedback seems also to be an important factor. As a matter of routine officers expect evaluations of their local office and subsequent follow-up. As noted earlier, feedback on how a local office has performed in meeting targets for the labor market is eagerly anticipated as a league table of performance. The follow-up needs to be sincere and seen to be acted upon. Of special interest is the importance that officers place on rehabilitation. This latter may be dependent on the special culture in this organization (see our earlier discussion on cadre organizations). Worthy of particular mention in this organizational climate are the key elements of mutual cooperation, behavior, and strong work ethic. Agency officers expect their manager to "actively contribute to ensure that people who have reported ill will return to work quickly." Here the agency's leaders have a special responsibility. Officers expect an active rehabilitation; if not, again they will be dissatisfied. However, they will be satisfied even if the rehabilitation is seen to work to only a small extent. Leaders also need to be seen to practice as they preach: rehabilitation is an area of critical importance for the agency's work in the labor market.

The agency's flat structure seems to contribute in no small part to the perception of officers of the workplace climate of which they are a part. Certainly the close proximity of officers in their workplace environments contributes to a climate where people are able to keep informed of the caseloads and case decisions of their colleagues. Other contributory factors are the processes of informal mentoring and—to a certain extent—the shared backgrounds in union work. In terms of formalized feedback, the 360° feedback mechanisms suggest that the perceptions of individual employees reflect in large part the workplace climatic conditions. We suggest that the positive benefits to workplace climate are the existence of feedback mechanisms that are perceived as open (visible, publicized, and a matter of routine), fair (conducted as a recognized part of regular performance assessment), and are seen to contribute to individual appraisal and office performance. In this case, the medium of encouraging feedback and the message that this conveys about participation and involvement are two faces of the same currency. The existence of these workplace practices influences the special nature of workplace climate in this cadre organization. An outcome is that individual perceptions combine into a whole that is larger than the constituent parts.

Mutual trust between agency officers and between the officers and their line managers is a critical component in the quality of service delivery. A lack of trust at grassroots levels (the officers who routinely deal with clients) would likely undermine efforts to develop an empowered organization. Employees might interpret moves by headquarters to exercise greater control as a lack of confidence by the central powers in employees' competencies in the regional branch offices. If this were to happen, it would deplete any gains in productivity from decentralizing powers to the regions. It might then be difficult to reconstitute a natural arena to stimulate employees of the cadre organization. If, at the same time, employees perceive other organizational changes that may be interpreted in a reduction in the quality of the work climate, there becomes a serious risk of a downward spiral in the quality of the workplace climate with resultant reductions in workplace productivity.

Endnotes

1. Emotional work (emotional labor) is highly skilled work that involves an employee dealing with highly emotional situations. Examples include people who work in the emergency services, in the professions of nursing and medical care, and in any industry with face-to-face or telephone contact with customers. People in emotional work need high levels of emotional intelligence (EQ) and expertise in managing social relationships and interactions often in stressful and often painful situations. A classic text on emotional work is

Arlie Russell Hochschild (2012 [1983]), *The Managed Heart: Commercialization of Human Feelings*, Berkeley: University of California Press. Also see discussions in Ronnie J. Steinberg and Deborah M. Figart (1999), Emotional Labor since *The Managed Heart*, *Annals of the American Academy of Political and Social Science*, 561, 8–26; Susan M. Kruml and Deanna Geddes (2000), Exploring the Dimensions of Emotional Labor: The Heart of Hochschild's Work, *Management Communication Quarterly*, 14, 8–14; S. Mann and J. Cowburn (2005), Emotional Labour and Stress within Mental Health Nursing, *Journal of Psychiatric and Mental Health Nursing*, 12(2), 154–162; Stephen Fineman (ed.) (2000), *Emotion in Organizations* (2nd ed.), London: Sage Publications; Alicia A. Grandey, James M. Diefendorff, and Deborah E. Rupp (eds.) (2013), *Emotional Labor in the 21st Century: Diverse Perspectives on Emotional Regulation at Work*, New York: Routledge Publishers.

2. See relevant discussions in Sean T. Lyons, Linda E. Duxbury, and Christopher A. Higgins (2006), A Comparison of the Values and Commitment of Private Sector, Public Sector, and Parapublic Sector Employees, *Public Administration Review*, July–August, pp. 605–618; Leonard Bright (2008), Does Public Service Motivation Really Make a Difference on the Job Satisfaction and Turnover Intentions of Public Employees? *American Review of Public Administration*, 38(2), 149–166; Wouter Vandenabeele (2011), Who Wants to Deliver Public Service? Do Institutional Antecedents of Public Service Motivation Provide an Answer? *Review of Public Personnel Administration*, 31, 1, 87–107; Leonard Bright (2013), Where Does Public Service Motivation Count the Most in Government Work Environments? A Preliminary Empirical Investigation and Hypothesis, *Public Personnel Management*, 42(1), 5–26.

3. See discussions in Barry Bozeman and Mary K. Feeney (2007), Toward a Useful Theory of Mentoring: A Conceptual Analysis and Critique, *Administration and Society*, 39(6), 719–739; Barry Bozeman and Mary K. Feeney (2009), Public Management Mentoring: What Affects Outcomes? *Journal of Public Administration Research and Theory*, 19(2), 427–452.

4. See Noreen Ritchie and Michael Connolly (1993), Mentoring in Public Sector Management: Confronting Accountability and Control, *Management Education and Development*, 24(2), 266–279; Brian J. Cooke (1998), Politics, Political Leadership, and Public Management, *Public Administration Review*, 58(3), 225–230; Richard Common (2004), Administrative Change in the Asia Pacific: Applying the Political Nexus Triad, *International Public Management Journal*, 7(3), 347–364; Sarini Saha (2011), City-Level Analysis of the Effect of Political Regimes on Public Good Provision, *Public Choice*, 147, 155–171; Belinda Luke, Kate Kearins, and Martie-Louise Verrynne (2011), The Risks and Returns of New Public Management: Political Business, *International Journal of Public Sector Management*, 24(4), 325–355.

5. Terence E. Deal and Allan A. Kennedy (1982), *Corporate Cultures: The Rites and Rituals of Corporate Life*, Reading, MA: Addison-Wesley, p. 49.

6. Edgar H. Schein (1997 [1992]), *Organizational Culture and Leadership*, San Francisco, CA: Jossey-Bass Publishers, Second edition. Especially see discussions in Chapter 8 (pp. 147–168) and Chapter 13 (pp. 254–275).

7. Geerte Hofstede (1991), *Cultures and Organizations: Software of the Mind*, London: McGraw-Hill, p. 179.

8. Brian Hunt and Toni Ivergård (2007), Workplace Climate and Workplace Efficiency: Learning from Performance Measurement in a Public Sector Cadre Organization, *Public Management Review*, 9(1), 27–47.

9. Joanne Martin and Caren Siehl (1983), Organizational Culture and Counter-culture: An Uneasy Balance, *Organizational Dynamics*, Autumn, pp. 52–64.

10. Guy S. Saffold (1988), Culture Traits, Strength and Organizational Performance: Moving Beyond "Strong" Culture, *Academy of Management Review*, 13(4), 546–557.

11. George C. Gordon (1991), Industrial Determinates of Organizational Culture, *Academy of Management Review*, 16(2), 396–415.

12. See relevant discussions in Alan L. Wilkins and William G. Ouchi (1983), Efficient Cultures: Exploring the Relationship between Culture and Organizational Performance, *Administrative Science Quarterly*, 28, 468–481; Cheri Ostroff, Yuhyung Shin, and Angelo Kinicki (2005), Multiple Perspectives of Congruence: Relations between Value Congruence and Employee Attitudes, *Journal of Organizational Behaviour*, 26, 591–623. Also see Brian T. Gregory, Stanley G. Harris, Achilles A. Armenakis, and Christopher L. Snook (2009), Organizational Culture and Effectiveness: A Study of Values, Attitudes, and Organizational Outcomes, *Journal of Business Research*, 62, 673–679.

13. Bo Rothstein (1998), *Just Institutions Matter: The Moral and Political Logic of the Universal Welfare State*, Cambridge: Cambridge University Press, especially p. 98ff.

14. See, for example, discussions in: Anne Marie Berg (2006), Transforming Public Services: Transforming the Public Servant? *International Journal of Public Sector Management*, 19, 6, 556–568; Ewan Ferlie (2007), Complex Organisations and Contemporary Public Sector Organisations, *International Public Management Review*, 10, 2, 153–165; Belinda Luke, Kate Kearins and Martie-Louise Verreynne (2011), The Risks and Returns of New Public Management: Political Business, *International Journal of Public Sector Management*, 24, 4, 325–355.

15. Bo Rothstein (1996), *The Social Democratic State: The Swedish Model and the Bureaucratic Model of Social Reforms*, Pittsburgh, PA: University of Pittsburgh Press, p. 51.

16. See discussions in Brian Hunt and Toni Ivergård (2007), Workplace Climate and Workplace Efficiency: Learning from Performance Measurement in a Public Sector Cadre Organization, *Public Management Review*, 9(1), 27–47.

17. Bo Rothstein (1998), *Just Institutions Matter: The Moral and Political Logic of the Universal Welfare State*, Cambridge: Cambridge University Press, p. 91.

18. See Henry Mintzberg (1991), The Effective Organization: Forces and Forms, *Sloan Management Review*, 32(2), 54–67; Robert Drago and Geoffrey K. Turnbull (1991), Competition and Cooperation in the Workplace, *Journal of Economic Behavior and Organization*, 15(3), 347–364.

19. See Henry Mintzberg (1991), The Effective Organization: Forces and Forms, *Sloan Management Review*, 32(2), 54–67.

20. On the public service ethic, see Gene A. Brewer, Sally Coleman Selden, and Rex L. Facer II (2000), Individual Conceptions of Public Service Motivation, *Public Administration Review*, 60(3), 254–264; Lotte Bøgh Andersen, Tor Eriksson, Nicolai Kristensen, and Lene Holm Pedersen (2012), Attracting Public Service Motivated Employees: How to Design Compensation Packages, *International Review of Administrative Sciences*, 78(4), 615–641.

21. See Kenneth Kernaghan (2007), *A Special Calling: Values, Ethics and Professional Public Service*, Studies and Discoveries Series, Canada Public Service Agency.

22. See discussions in Lotte Bøgh Andersen, Thomas Pallesen, and Heidi Houlberg Salomonsen (2013), Doing Good for Others and/or Society? The Relationships between Public Service Motivation, User Orientation and

University Grading, *Scandinavian Journal of Public Administration*, 17(3), 23–44; Lotte Bøgh Andersen, Torben Beck Jørgensen, Anne Mette Kjeldsen, Lene Holm Pedersen, and Karsten Vrangbæk (2013), Public Values and Public Service Motivation: Conceptual and Empirical Relationships, *American Review of Public Administration*, 43(3), 292–311; Rachel Gabel-Shemueli and Ben Capell (2013), Public Sector Values: Between the Real and the Ideal, *Cross Cultural Management*, 20(4), 586–606.

23. For discussions on the relationship between employee attributes and attitudes and public sector work, see Sangmook Kim (2012), Does Person-Organization Fit Matter in the Public Sector? Testing the Mediating Effect of Person-Organization Fit in the Relationship between Public Service Motivation and Work Attitudes, *Public Administration Review*, 72(6), 830–840; Anne Mette Kjeldsen and Christian Bøtcher Jacobsen (2012), Public Service Motivation and Employment Sector: Attraction or Socialization? *Journal of Public Administration Research and Theory*, 23, 899–926; Anne Mette Kjeldsen and Lotte Bøgh Andersen (2013), How Pro-Social Motivation Affects Job Satisfaction: An International Analysis of Countries with Different Welfare State Regimes, *Scandinavian Political Studies*, 36(2), 153–176.

24. Discussions of this issue can be found in Carolyn J. Heinrich (2002), Outcomes-Based Performance Management in the Public Sector: Implications for Government Accountability and Effectiveness, *Public Administration Review*, 62(6), 712–725; Sven Modell (2005), Performance Management in the Public Sector: Past Experiences, Current Practices and Future Challenges, *Australian Accounting Review*, 15(3), 56–66; Frank H.M. Verbeeten (2008), Performance Management Practices in Public Sector Organizations, *Accounting, Auditing and Accountability Journal*, 21(3), 427–454; Suwit Srimai, Nitarath Damsaman, and Sirilak Bangcholdee (2011), Performance Measurement, Organizational Learning and Strategic Alignment, *Measuring Business Excellence*, 15(2), 57–69.

25. Focus on quality improving quality.

26. See discussions in Chris Argyris (1977), Double Loop Learning in Organizations, *Harvard Business Review*, 55(5), 115–125; Chris Argyris and Donald Schön (1996), *Organizational Learning II: Theory, Method and Practice*, Reading, MA: Addison-Wesley Publishing.

27. Caroline J. Heinrich (1999), Do Government Bureaucrats Make Effective Use of Performance Management Information? *Journal of Public Administration Research and Theory*, 9(3), 363–393.

28. Carolyn J. Heinrich (2002), Outcomes-Based Performance Management in the Public Sector: Implications for Government Accountability and Effectiveness, *Public Administration Review*, 62(6), 712–725.

29. Caroline J. Heinrich (1999), Do Government Bureaucrats Make Effective Use of Performance Management Information? *Journal of Public Administration Research and Theory*, 9(3), 363–393.

30. Carol Propper and Deborah Wilson (2003), The Use and Usefulness of Performance Measures in the Public Sector, *Oxford Review of Economic Policy*, 19(2), 250–264.

31. A. Dixit (1997), Power of Incentives in Private versus Public Organizations, *American Economic Review*, 87(2), 378–382; Avinash Dixit (2002), Incentives and Organizations in the Public Sector: An Interpretive View, *Journal of Human Resources*, 37(4), 696–727.

10

Designing a Service Dream: Excellence from Merging Public and Private Service

In this chapter we present two case studies: one each from the private and public sectors. We have based these case studies on our respective experiences in these types of organizations. We have combined details from several organizations into each case study description. We see merit in blending together attributes of each organization for the purposes of analysis and discussion. The details are from real, living organizations that exist and thrive. We detail key features of each type of organization and discuss the benefits for designing service-driven organizations. Our descriptions are multifocused. We describe and discuss detailed attributes of organizational structures, human resources (HR), and employee competence development and leadership. We meld each of these features into frameworks for organizational development. At the heart of our descriptions are key attributes that, in combination, could form the design of a new type of organization. This organization would incorporate the best of both worlds. As such, the organization's managers, executives, and staff would be well placed to shape the organization's future by mobilizing, focusing, and aligning internal resources (physical capital, financing, and human capital, including explicit and implicit competencies). In concluding, we share our reflections on this new organization, the challenges it will face in an increasingly technological world from an operational standpoint, and the ways in which service design will enable organizations not only to meet these challenges, but also to continually reshape themselves for futures unknown, but certain to be turbulent.

Excellence in a Private and in a Public Organization

Below we describe three hypothetical organizations. One is a private sector organization, and the second a public sector organization. The third is an amalgam of the two. We have developed our ideas over time from many different locations and environments in which we have been privileged to work. In our careers we have worked in countries on several continents. We therefore draw on many different organizational and national cultures in which we have enjoyed working. These organizations might exist somewhere in the real world—at least we would certainly like to think so. In our current context they are based on our own personal experiences over a number of decades with several dozen public and private sector organizations. Readers who feel that they recognize specific organizations are welcome to conjecture, but are likely to be wrong.

We have combined the two organizations into a third hybrid organization to represent a Socratic ideal of private-public organizations. We are certain that the optimal organization of the future needs to learn from excellence in both private *and* public organizations. In our experience, excellence is not the preserve of one type of organization. Certainly, each type of organization can offer valuable lessons in organizational structure and design, workplace practices, and management.

We first set out key attributes of each of the private and public sector organizations as a form of case study. As mentioned, these case studies are based on real, living organizations of which we have had direct experience over our working lives. We then discuss the traditional understanding of the strengths of each type of organization. As a way of making sense of the different organizations, we also discuss ways in which we can learn from the strengths of each sector organization and implement a combination of these strengths in the hybrid organization. Our intention is to construct an organization using the best of the private and public sector worlds. We conclude by discussing how to optimize the learning from the private and public organizations as inputs for the design of new types of organizations.

Service Excellence in a Large International Private Sector Organization

Our private sector organization is a company whose operations are mainly in the area of logistics. The organization is rather large, has a global reach, and has tens of thousands of employees. Although its operations span the world, its main activity is in Europe, where it is headquartered. The organization has been in existence for many decades. During this time it has invariably been successful, although it has seen peaks and troughs of profitability. Over a period of about a decade, it was extremely successful in its business and achieved leadership in its industry as an exemplar of business excellence. Competitors tried to emulate its success. The organization has focused very strongly on providing very high levels of service for a segment of its customers who could (and were happy to) pay for added-value services. The organization also tried to provide a good basic range of services for all of its customers. In common with many large organizations, it has in the past been overloaded by a rather heavy bureaucratic structure. As a partial solution to this issue, the organization's executives believed strongly in creating business innovations. In the main, these initiatives were eye-catching and successful and competitors tried to replicate them.

A key area that received priority of management time and focus was the frontline (customer-facing) employees. The organization tried to reduce the number of formal rules and regulations that had built up over time so that frontline staff would become (of necessity) more empowered to make decisions without recourse to immediate line managers. The organization also aimed to break down its rather hierarchical structure, which was a corollary of its time-honored traditional bureaucracy. Executives and managers also tried to reshape the organization into a flat, low-level organizational structure. Reducing levels of reporting and spans of control was seen as a means to reform the perceptions of employees whose routines and daily work tasks focused on servicing customer needs. To this end an overlay matrix was designed and introduced in order to activate all functions. This matrix was used throughout the organization and affected all functional units, regardless of whether these were frontline (customer-facing) or traditional support services (such as information and communications technology (ICT) or HR, such as recruitment, training, and payroll services). Introducing the matrix as a way of representing and publicizing responsibilities helped ensure that employees were aware of their place in the organization's structure and operations. The overarching aim of the new matrix was to aid workplace clarity.

A key and dominant feature of this service organization was its highly charismatic leadership. The leader was personable and able to speak to everyone, and made a point of doing so. This is usually referred to as having the common touch. The leader was high profile and an effective communicator. This leadership style generated a widespread feeling by employees of engagement, involvement, and participation. However, in reality, although this represented a change from traditional leadership styles, it was still not direct participation in a strictly democratic sense. The command structure remained top-down and activities were invariably initiated and driven by the top leader. That there was now a chance of increased dialogue between the workforce and its leader was a subtle difference over an autocratic style of leadership.

The focus of the leader and his executive team was clearly on the real and prime customers—real in the sense that the organization knew the profiles of its customer base and prime in that data generated by revenues figures clearly showed that around 20 percent of customers generated a high share of revenues and profits.[1] It was therefore expected that frontline (customer-facing) personnel had priority. However, at the same time, the organization emphasized the importance of good internal service to internal customers, i.e., support functions and in-house services. Employees were constantly reminded that this too was a priority area. Of particular importance was the leadership at all levels and in all departments. This feature of organizational development emphasized the responsibility and roles of mid-ranking leaders (e.g., department heads and unit supervisors) in providing service to employees in their units. In order to empower employees, and in particular frontline employees who were in daily contact with customers, there was a drastic reduction of rules and regulations. Rule books and lengthy guidelines were purposefully discarded. Thereby, frontline personnel gained more freedom to give the best possible service to the customer in every situation. Devolving decision making to frontline employees is a key component of service excellence. The award-winning international health resort and spa Chiva Som (Haven of Life) trains its service employees in a can-do attitude that in practice means that the resort's healthcare professionals, therapists, and employees aim to do the best they can do for the guest.[2]

In our case study organization it was emphasized that every meeting of frontline personnel and the customer was a unique situation, and one that would shape the customer's perception of the service. Success at this moment of truth received the highest possible priority. The success of this moment of truth had to be understood by everyone throughout this rather large organization. The objective was that each and every one of the tens of thousands of employees should be able to internalize the concept of the moment of truth. The rather high level of success in this objective was very much dependent upon the strong charismatic personality of the organization's leadership. The concept of alignment of all its employees became very important and, to a large extent, was fulfilled. This form of top-down creation of alignment can be looked at in relation to the concept of a cadre organization.[3]

In the main, the organization was driven by top-down initiatives in terms of strategic direction and ambition. However, in practice it had many features that are associated with a more participative and democratic organization in terms of workplace practices. One example was a stated aim to move toward a flat organization structure by removing layers of control, supervision, and reporting. Another example was deregulation (removing what some employees saw as petty bureaucratic rules) to give more freedom to individual initiative and encourage personal creativity. Internal deregulation also facilitated stronger frontline staff-customer relationships.

In addition, the organization introduced an elaborate follow-up and evaluation system of the organizational functions at all levels. Particularly, the leadership was evaluated in a 360° feedback process. These evaluations (in effect, personnel/leadership audits) were conducted throughout the organization and used at all levels of responsibility to create action plans for improvements. To a large extent, the focus was on leaders' success or difficulties in handling human relations. This was both internal (pertaining to staff and work teams) and external (relating to relationships with customers). Initially, customer relationships and personnel were separated into two types of internal audits. Over time, these two processes were merged to provide a holistic picture. These rather elaborate evaluations/audits combined with plans of actions proved to be a very successful method of improving the function of this large organization. Compared to traditional quality assurance (QA) types of studies, which may often become one-off exercises, this process became a vital and ongoing part of a new kind of learning organization. It created a mutually supportive human relations (leadership personnel) process of organizational learning. This is somewhat rare. From year to year could be seen numerous examples of dramatic improvements in leadership skills and leadership outputs. In fact, experiences from these processes might call into question many traditional programs of leadership development. The types of audit put in place by this organization highlighted the development of each organizational unit. Each organizational unit consists of a unique set of individuals and their communal interactions. Learning has to be mainly based on this mutual process between unique individuals. This will always be different from training leaders at off-site events (e.g., so-called away-day seminars) and separated from their daily social milieu and workplace context of socialization. However, to be successful in the longer term, these types of processes need to be allowed to continue over prolonged periods of time. In this way, there will be an organizational context for the learning (rather than what might be termed learning in a vacuum). This will not occur if the process changes in the short term (e.g., on an annual basis). To ensure change momentum and robustness of learning, the processes need to be followed up over shorter-term periods (such as on a 6-month cycle). The dynamics of this organizational learning have some similarities to some of the concepts developed by the Tavistock Institute in its learning conferences.[4]

Year on year over a period of ten years, this private organization was very successful. Undoubtedly, one factor of this success was the quality of high-profile charismatic leadership throughout the organization. Most importantly, this leadership underscored the short-term success of the organization. However, a valuable supportive role was played by follow-ups and evaluations set in place by these leaders and used as a tool for a continuous process of organizational learning. This helped achieve sustainable success over a longer timeframe.

The top leadership and especially ongoing dialogue between top leaders and frontline personnel were important features in creating the driving force for continuous change and organizational evolvement. But at the same time, this also created a number of difficulties. Many high-level leaders experienced a loss of power and influence from what they had previously enjoyed. Many of these leaders were also actively critical of the evaluation processes as a tool for organizational and human resources (HR) development. Some high-level leaders believed that these types of processes represented a personal threat to their self-esteem, on both personal and professional levels. At times, this also created conflict between the frontline personnel and some high-level leaders. However, in the long term, such conflict can be avoided if the process of organizational learning can achieve an understanding that (in some cases) changes are allowed a longer timeframe. This requires a detailed examination of the issue on a case-by-case basis coupled with an acceptance by all concerned parties that local branches of the organization (i.e., subunits or outstations of the organization) are allowed to solve their own operational problems (on their own initiative) in cooperation with other parts of the wider organization.

In any organization there are issues of succession and continuity. In this private organization when the creative and innovative leadership left to undertake other missions, some of the impetus and momentum also disappeared from this rather unique organization at a critical stage in a process of change. This notwithstanding, valuable lessons can be learned and used by other organizations in both the private and public sectors. Modern information and communication technology (ICT) can play a much larger role in this process of change. Through judicious use of ICT, organizations could make processes of auditing and organizational change into a fully decentralized process for each of their work subunits. This would then be a form of organic self-controlled, self-directed organizational learning. In practice, the employees and their leader in an organizational subunit (such as an operating division or functional department) would manage their own process of auditing, making department-specific plans of actions within the framework of the organization's overall vision and missions. The department would also be responsible for implementing the necessary actions. This would include control of its own economic balance sheet.

As a multinational corporation (MNC), this organization operated in many different parts of the world. As such, it was exposed to the influences of many different national cultures. Initially, some employees questioned whether the types of methods put in place in Europe could be successfully transposed in many different types of cultural contexts. There is much here to describe and discuss, but to do so fully would require another book. In summary, the methods were applied without any major hiccups in all different countries involved. Naturally, while there was some adjustment to conform to local norms and expectations, the processes retained the key features of those developed at the European headquarters. Organizations often contain several different (and often conflicting) professional cultures.[5] In this organization it was evident that the differences between individuals and local work units were much larger than the differences between units from different cultural areas. As an example, the differences in worldview and mind-set between, say, engineers and marketing and sales personnel (irrespective of geographic locale) were much larger than the differences of the people working in different geographic regions. Bridging the professional divide was more important than spanning the cultural gap.

Service Excellence in a Large Public Sector Organization

In recent years public service has undergone a shift in emphasis with a stronger focus on process effectiveness leading to mechanisms to deliver good service to the nation at large. The public sector organization described here is proud of its long heritage and its proven service to the monarch and nation. This public sector organization claims its roots back several centuries, to the beginning of the development of the nation-state. In many countries (such as France, the UK, Sweden, China, and Japan) public service was traditionally to the crown, and subsequently to the nation-state.[6] In this model the emphasis is less on *public* service (i.e., to the civil population) and more on service provided to authority figures within the state and to matters under their bailiwick (such as taxation and mobilizing armies). In this model of public service, the organization takes a pyramidal shape.

During the mid- to late nineteenth century in our case study example the public administration saw many reforms as the nation continued to develop and tackle the not insubstantial challenges of modernization. Reforms over several decades gave shape, structure, and functions to the nascent public administrative infrastructure. The present-day organization, structural configuration, and work ethos owe much to the many changes initiated in these reforms. Development of the public sector organization continued apace throughout the twentieth century. And although there have been periods of social and political upheaval, there have also been consistent programs of reforms introduced by politicians of different hues. The latter decades of the twentieth century and the first decade of the twenty-first century have seen further reform effort, much of this initiated within frameworks of new paradigms for public sector management.

In the early 1990s the book *Reinventing Government* delineated new public management (NPM).[7] NPM imported ideas from private sector organizations with a view to transforming public organizations to resemble more closely the management styles, functionalities, and operational efficiencies of commercial enterprises. To this end, NPM envisaged refocusing public management effort in public service delivery from bureaucratic rigidity to cost-effectiveness. The former situation saw rule-based processes of service. Standardized (supply-driven) services were the norm, and work procedures tended to be overly bureaucratic and cumbersome with an emphasis on past precedent. The focus and content of public services has more often been decreed by politicians and ministers, rather than shaped by the civic needs of citizens.

In our case study organization, the past few years have seen reforms focused on developing the sector's human resources (HR) competencies. The intention is to reshape the organization and its prevailing ethos to development, and reform the constituent organizations from a traditional model to one more focused on effective delivery of public services. In essence, this is a move from top-down organizational development toward initiatives generated from the bottom up. In common with its neighbors, the country in which this organization is situated was affected by the 1997 Asian financial crisis. In its aftermath, there were further reforms of the public administration, especially in articles contained in a new constitution. These reforms encompassed more public participation in the mechanisms of governance, including public debates of political and social issues affecting the nation's citizens. For example, the subsequent national development plan was composed with reference to public forums conducted nationally.

The past few years have seen reforms to the organization's infrastructure, in the main intended to push forward modernization (in part to comply with NPM principles, in part to engage a new type of career civil servant, and in part to address the needs of a developing society). Key emphases are to reinforce decentralization of the mechanisms of public governance, ensure that there are robust processes for engagement of the nation's citizens in civil and social debate, and move closer to an optimal size of the public sector to be able to deliver timely public services (e.g., education, healthcare, care of an aging population). The renewed focus is effectiveness of resource use, and timely delivery of public services as needed (i.e., public services that are demand driven rather than supply driven).

Not surprisingly, this focus is driving changes in the workplace culture and philosophy in the public bureaucracy. The focus of resources is now to be driven by demands of public service needs. Organizational restructuring to achieve effectiveness of limited resources is beginning to replace the former emphasis on organizing resources for efficiency. In essence, effectiveness means *doing the right things* (in this case, serving society), while efficiency *means doing things right* (following rules and regulations). Underpinning each of these focuses is the reformation of organizational structure from a busy but unproductive bureaucracy to a service-oriented organization. This involves upending the shape of the organization from a strict hierarchy to one focusing on the competencies and service abilities of public servants. In practice, this means that rational decisions for the deployment of resources for public services by officials who can adapt to changing social needs replace knee-jerk reactions to rule-governed preplanned outputs. In the NPM model of public service, key people skills include flexibility (of both thinking and actions), empowerment, monitored performance outputs (transparency and accountability), and an ability to make decisions in team-based situations. This is a break with past thinking and behaviors. Whereas before the organization looked inward for resources and to the past for precedents of planning and resource allocation, it now looks outward to communities for appropriate levels of resources and to planning for the future in relation to timely data from national and local statistics.

Learning between Private and Public Sector Organizations

In many parts of the world, the public sector is looking with great interest at the private sector to learn about best practices. Examples of areas for learning are suggested to include organization structure, project costing and budgeting, and management (including decision making and practices of personnel management). In many ways this is a gross mistake. The conditions under which private sector organizations operate are different from public sector organizations. In particular, in Scandinavia and throughout Europe, public organizations have played both copycat and catch-up. This can be a dangerous error of judgment.

In this chapter we have reflected on our own experiences of working in and being involved with public organizations. Our description of this very large public organization is an amalgam of many organizations of this type. We have been working in, employed by, and have conducted research in this type of organization in Europe, Asia (particularly Southeast Asia), and the southern parts of Africa.

Based on our experiences the learning between private and public sector organizations ought to be mutual and (at least) bidirectional. It is a necessity for public sector organizations to look for the best in the private sector that are suited to be applied in specific types of public sector. The key word here is *suited*. Not all public organizations are, on a broad basis, suited to learn from the private sector, more than from specific examples. Typical areas that it is difficult to generalize from private to public sector work include the law and legal matters, for example: issues involved in policing, prison services (correctional facilities in general), and similar public service organizations. Conversely, public enterprises such as electricity boards, water companies, railway systems, public transport, telecommunications, and similar public services offer possibilities to learn from the private sector. It is not unusual for services in these areas to be delivered by private enterprises, and even for public services in these areas to become privatized. Obviously, all moves to make public organizations fully private must be assessed and conducted with an understanding of the great importance of infrastructure requirements of a nation, province, or local community. To ignore this essential condition would be to risk reverting to a supply-driven model of public services.

Of course, infrastructure requirement needs of the public sphere can be met in many different ways. Theoretically, it could be relatively straightforward for a fully government-owned organization to arrange a fully public organization to meet the requirement needs of a public infrastructure. Widespread experiences (sometimes horror stories) of a public organization's inability to meet the public needs of infrastructure are not primarily a problem of public or private ownership. Rather, this is more a matter of inefficient management and decision making and (mis)allocation of available resources. In turn, this is in many cases related to a more formal decision-making structure due to general public concerns and formal legislations. What we have learned from Chapter 9 is that some private organizations are able to organize very flexible and decentralized decision-making processes. These processes are often designed to facilitate a decentralized process of learning and development. This type of organization can lead to a more decentralized faster decision making without losing core governance mechanisms and control by those in authority, such as a board of directors. One would perhaps expect good governance to be of a higher quality in the public sector than in the private sector due to concerns about the "public purse" and the need for openness and transparency. This is not necessarily so.

One additional reason for the interest from the public sector to learn from the private sector is the spread of the market economy in almost all countries around the world. Recent history suggests that command economy principles have been heavily marginalized and find their adherents in a relatively small number of nations around the world. It is not always obvious or a fact that privatization of a public organization will lead to increased efficiencies and lower operational costs. In many cases it may be the opposite. In the past few decades privatization has taken place in many areas of previously public organizations. An interesting example is the health sector. Many cases indicate that publicly owned organizations in the health sector are more cost-effective than in the private sector ownership. We've each had opportunities to see firsthand examples of how public organizations in certain areas of healthcare can be improved by the learning from organizations in similar areas of the private sector.

In the public sector, preparedness for change varies to a very large extent. In many areas of the public sector organizational cultures inherently resist the process of change. This is often a facet of the organizational culture, as such, but is also many times an inherent feature of employees' expectations of their organization ("change is not good and should be resisted"). On the other hand, in many types of public organizations it might be relatively easy to make changes. This is the case in what we call cadre organizations.[8] A number of distinctive features define a cadre organization. Its members tend to share an ideology, there is often a lack of formal levers of control (members organize, direct, and manage themselves), and there is often a strong sense of mutual commitment.[9] Whereas in many forms of organization the mission is set from the top down (i.e., from senior executives), in a cadre organization the members themselves exhibit a collective (mutually supporting) sense of the organization's beliefs, guiding principles, mission focus, and ways in which the organization should be conducting itself. Cadre organizations display a strong sense of purpose and high levels of consensual behavior. Notions such as duty, service, and helping others are among the key propositions in a cadre organization. For these reasons, cadre organizations can be found in political parties, unions and professional bodies, religious groups, and charitable organizations. The characteristics of a cadre organization make it very easy to implement changes, as long as changes are in line with the guiding mission of the organization as understood by the membership.

The specific characterization of cadre organizations might, in turn, make them resistant to change if proposed changes do not accord with what the members feel are the guiding principles and their organization's stated mission. Typical examples of cadre organizations can be found in some parts of the Swedish government bureaucracy.[10] Strong organizational cultures of employee solidarity can also be found in many government organizations and private organizations.[11] In Thailand, for example, public sector officials are designated as servants of the monarch. Naturally, this can create a situation where employees have a very strong attachment to tradition and a tendency to resist change, especially if this is to threaten time-honored ways of working.[12] Privatization of public enterprises or part of government organizations is nearly nonexistent in spite of a number of attempts to privatize public enterprises. Evidently, public and private organizations can learn from each other. Most likely, public organizations can learn from *good* examples from the private sector. In the long term, public organizations could probably benefit more than the private sector in creating decentralized learning organizations.

Endnotes

1. The Pareto effect (or Pareto principle), also known as the 80:20 rule, was conceived by Joseph M. Juran (1904–2008) and named after the Italian economist Vilfredo Pareto (1848–1923). The Pareto effect states that in any phenomenon, 20 percent of causes generate 80 percent of effects (such as that 20 percent of customers generate 80 percent of sales revenue). See discussions in Alan J. Dubinsky and Richard W. Hansen (1982), Improving Marketing Productivity: The 80/20 Principle Revisited, *California Management Review*, 25(1), 96–105; Bill Birnbaum (2004), Use a Pareto Diagram to Develop Strategy, *Consulting to Management*, 15(1), 15–16.
2. See relevant sections in Marc Cohen and Gerard Bodeker (2008), *Understanding the Global Spa Industry: Spa Management*, Oxford: Butterworth-Heinemann.
3. In essence, a cadre organization can be seen in mission- and ideology-driven organizations such as charities, religious organizations, and political parties. Such organizations tend to be non-profit seeking (at least in principle).
4. The Tavistock Institute was founded in 1947 as a not-for-profit organization that "applies social science to contemporary issues and problems" (see http://www.tavinstitute.org). For descriptions of the work of the Tavistock Institute, see, for example, Richard K. Brown (1967), Research and Consultancy in Industrial Enterprises: A Review of the Contribution of the Tavistock Institute of Human Relations to the Development of Industrial Sociology, *Sociology*, 1(1), 33–60; Eric Trist and Hugh Murray (1997), Historical Overview: The Foundation and Development of the Tavistock Institute to 1989, in E. Trist, F. Emery, and H. Murray (eds.), *The Social Engagement of Social Science: The Socio-Ecological Perspective* (vol. III), Philadelphia: University of Pennsylvania Press, pp. 1–35.
5. See, for example, Edgar H. Schein (1996), Three Cultures of Management: The Key to Organizational Learning, *Sloan Management Review*, 38(1), 9–20. Also see Edgar H. Schein (1996), Culture: The Missing Concept in Organizational

Studies, *Administrative Science Quarterly*, 41(1), 9–20; Harrison M. Trice (1993), *Occupational Subcultures in the Workplace*, Ithaca, NY: Cornell University, ILR Press, especially Chapters 1–3.

6. See discussions in Anthony Kirk-Greene (1999), *Crown Service: A History of HM Colonial and Overseas Civil Services 1837–1997*, London: I.B. Tauris and Co.; Benjamin A. Elman (2000), *A Cultural History of Civil Examinations in Late Imperial China*, Berkeley: University of California Press; Pan S. Kim (2002), Civil Service Reform in Japan and Korea: Toward Competiveness and Competency, *International Review of Administrative Sciences*, 68, 389–403.

7. David Osborne and Ted Gaebler (1992), *Reinventing Government: How the Entrepreneurial Spirit Is Transforming the Public Sector*, Reading, MA: Addison-Wesley.

8. Brian Hunt and Toni Ivergård (2007), Workplace Climate and Workplace Efficiency: Learning from Performance Measurement in a Public Sector Cadre Organization, *Public Management Review*, 9(1), 27–47.

9. For fuller descriptions see Bo Rothstein (1996), *The Social Democratic State: The Swedish Model and the Bureaucratic Model of Social Reforms*, Pittsburgh, PA: University of Pittsburgh Press; Bo Rothstein (1998), *Just Institutions Matter: The Moral and Political Logic of the Universal Welfare State*, Cambridge: Cambridge University Press; Brian Hunt and Toni Ivergård (2007), Workplace Climate and Workplace Efficiency: Learning from Performance Measurement in a Public Sector Cadre Organization, *Public Management Review*, 9(1), 27–47.

10. Brian Hunt and Toni Ivergård (2007), Workplace Climate and Workplace Efficiency: Learning from Performance Measurement in a Public Sector Cadre Organization, *Public Management Review*, 9(1), 27–47.

11. See, for example, discussions in Daniel J. Caron and David Giauque (2006), Civil Servant Identity at the Crossroads: New Challenges for Public Administrations, *International Journal of Public Sector Management*, 19(6), 543–555.

12. See, for example, Suntaree Komin (1990), Culture and Work-Related Values in Thai Organization, *International Journal of Psychology*, 25, 681–704; Sirirat Choonhaklai and Pangorn Singsuriya (2008), Thailand's Approach to Achieving Effective Leadership: Culture and Outcomes, *International Employment Relations Review*, 14(2), 38–55.

11

Service and Technology in Retailing: History, Concepts, and Concerns

Retailing and the Moment of Truth

For centuries, retailing comprised direct face-to-face contact between a service provider and the customer. In traditional models of retail service encounters, face-to-face transactions predominated. Such encounters were composed of moments of truth between the service employee and the customer. This sociocommercial relationship meant that the retail service encounter was rich in moments of truth. But the situation didn't stop there. Retail outlets also needed to develop business and often personal relationships with producers and their suppliers, including any necessary middlemen whose role was to ensure smooth transactional flows. In general there was a strict demarcation of roles within the supply chain from producer to retailer. Transportation of produce was in some cases the preserve of the producer. Storage and packaging was undertaken by the retailer. Needless to say, there were exceptions: a retailer might use its delivery vehicle to collect produce from a centralized facility such as a wholesale market or a warehouse. The market would then be a link in the supply chain with the responsibility for dividing large quantities of produce into smaller portions (from wholesale batches to individual retail qualities). When produce was especially bulky or required preservation (such as refrigeration), bulk storage or refrigeration facilities might be sited in areas where comparatively lower rents allowed larger premises. The invention of refrigerated vehicles allowed this type of facility to become mobile and shifted the task to another link in the logistical supply chain: the specialized transportation company. The increased size of such vehicles meant that deliveries could be made overnight in time for the next business day. Thus, packaging that was sized for an individual customer (such as wrapping meat, cheese, or fish) and done on a counter in the customer's presence shifted earlier in the supply chain toward the producer. A number of retailer-customer moments of truth are lost. These are replaced by convenience for both retailer and customer. Convenience, such as speed of service or product delivery, low price, packaging, or ready availability of the product, tends to be a ready substitute for detailed and personal service. Personal service in the traditional sense tends to be expensive (and often part of a luxury service offering). Innovations and novelty in retailing began to be seen in the mid- to late 1800s. The rise of the department store is attributed to mass migrations across oceans and from the countryside to urban centers, which led to cities becoming more populous during this period.[1]

The Birth of Modern-Day Retailing: Le Bon Marché

Toward the end of the third decade of the nineteenth century Aristide Boucicaut (1810–1877) opened a general store in Paris. Called Le Bon Marché (The Good Deal), the venture was successful and the entrepreneur wished to expand. Around the mid-nineteenth century, Boucicaut commissioned architect Louis Auguste Boileau to propose a purpose-designed building in Paris to house a much bigger store. Le Bon Marché held a number of new experiences for its customers. In contrast to other contemporary venues for buying goods, the store had elegantly designed premises and spacious shopping areas. The merchandise was displayed on open counters rather than in locked cabinets effectively guarded by the shop assistants. The store offered merchandise at fixed prices (indicated on price tags) and a guarantee of replacement if the merchandise later proved unsuitable. Clearly these were key attributes for business success. Several years later, the store expanded further when the engineer Monsieur Gustave Eiffel was approached to design a suitable building in which to set out his vast range of merchandise. Monsieur Eiffel (later to create the famed Eiffel Tower for the 1889 World's Fair in Paris) designed a department store that complemented the elaborate displays of exotic fabrics, clothing, and accessories. With countertops used to display merchandise, shop assistants and customers occupied the same space on the sales floor. Proximity of server and served replaced distance across a wooden counter, and instead of being handed goods, customers handled the merchandise they wished to buy (or merely to touch). Customers' spending appetites were stimulated by tactile contact with their purchases. Customers, who tended to be members of the burgeoning affluent middle classes or merely those with middle-class aspirations, flocked to the new concept in retailing and service. Personal service intimacy was an overnight success. By 1914 Le Bon Marché was the biggest department store in the world. Around the same period, Lord & Taylor and Macy's in New York and Marshall Field's in Chicago were establishing their reputations with similar service offerings. Slightly later, in the 1880s, Grace Brothers, David Jones, and Myers were developing their retailing businesses in Australia.

Some decades later, in the early 1900s, American retailing entrepreneur Harry Gordon Selfridge (1858–1947) developed his retailing concepts and knowledge in Chicago. Following a holiday visit to London in 1906, Selfridge decided to invest in his own department store in that city. Conjecturing that in terms of size or range of merchandise he could not compete with similar stores in the large cities of America or in Paris, he sought ways to differentiate his store. Clearly the store would have to be located in London. At that time London had a population of about 6.5 million and was the administrative and financial center of an empire "on which the sun never sets," although this was waning and would never reach the imperial heights of the mid-Victorian era. In 1909 Selfridge opened his emporium on London's Oxford Street. Selfridges was not London's first department store. That honor belonged to the store owned by William Whitely, a seller of drapery and linens who had opened his shop in Bayswater in West London. On November 5, 1876, just after Whitely had expanded his drapery business to include departments selling fresh meat and groceries, local traders formed a mob to protest this outrage. A sizable mob gathered in Westbourne Grove and cast an effigy of William Whitely into a bonfire set up to commemorate Guy Fawkes' night.[2]

Over time, Selfridges combined merchandise with attractions to draw in customers. A main attraction was the breadth of merchandise offered in the store. Before the inception of the department store, a key to success for a merchant was specialization.[3] Most shops focused on one sort of good and tended to congregate in particular areas. In London, Smithfield was the meat market, Covent Garden the market for fruit and vegetables, and Billingsgate specialized in fish. Craftspeople also tended to bunch together so that Savile Row is still famous for men's suits, and nearby Jermyn Street has numerous shirt shops.

The contribution of these early pioneers to the modern shopping service experience was to allow customers the sensory experience of handling the merchandise that previously had been placed out of customers' reach and handled solely by the shop assistants (often wearing cotton gloves).[4] Stores were designed to be what is now called open plan to give customers a sense of space. In turn, this encouraged customers to see shopping as a leisure activity, rather than as a household chore delegated to the work of servants and housemaids. Within the stores themselves were nonshopping venues (such as rooms in which to relax, libraries perhaps, and quiet rooms for reading and writing). Furnishing in these spaces tended to be luxurious, which added to the cachet of frequenting the store. In times when streets were not wholly paved or free of refuse, the attractions of space, tranquility, and soft furnishings seemed to be an irresistible combination. The store owners and of course their designers began a trend that endures until today. In the selling zones of the store, shop assistants were hired for their ability to sell in a subtle manner—not too pushy, but providing sufficient information about the merchandise to whet the customer's appetite to try, to taste, to purchase. Moreover, Le Bon Marché also offered customers a guarantee of replacement if the merchandise later proved unsuitable. As an amalgam of experiences, coupled with the sensory perceptions (especially touch, taste, feel, and smell) of the merchandise, the new shopping experience must have seemed extraordinary. So much so that it was a template for other industry sectors. Almost simultaneously, world-class hotels were developing models of service delivery for their clientele.

What the Retail Pioneers Gave to Modern-Day Retailing

At Le Bon Marché, the shop assistants stood in close proximity to the customers. This facilitated a more personal style of service, as the assistant could now allow or even encourage the customer to touch the merchandise. The lack of a physical distance and wooden counter to act as a barrier and goods on open display encouraged customers to feel the merchandise with or without the permission of the shop assistant. While this was far from self-service, it was a radical departure from the structure of giving service to customers, which continued to be the norm in shops, including Le Bon Marché's competitors. Formerly, a wooden or glass counter separated the customer from the sales assistant. Customers mainly approached a sales assistant when requesting service (such as information on price, quality, or availability) or when intending to purchase merchandise. Merchandise was visible but out of reach in display cabinets, in shop windows, or was placed some distance above the shop floor. Now with a closer proximity of the customer to the merchandise on open display, the customer is given more choices. She not only is tempted to touch and feel the goods, but also can handle the goods while discussing their qualities with the sales assistant. As is now recognized, customers who touch products are more likely to spend more than customers who do not touch products.[5] It would be *faux naïf* to believe that M. Boucicaut was unaware of this when he commissioned the design of his emporium.

Psychologists say that the sense of touch is the most emotive of the human senses.[6] As infants our sense of touch is the first of our senses to be developed and provides our main engagement with our environment. Touch is said to be a proximal sense, as the physicality of the tactile sensation is not moderated or filtered by air (as in the senses of sight, hearing, and smell) or by saliva (in the sense of taste).[7] The main facilitator of our tactile sense is the skin. Sight-impaired people learn to navigate their environment using touch as a key sense, perhaps using a stick or cane and often complemented by a sense of hearing more attuned to their immediate environment. The skin is the oldest and largest of the human sense organs.[8] It covers our skeletal and muscular form like a flexible cloak acting as a means of communication and protecting us like a shield.[9] It has the remarkable facility to renew.

The average adult human male has about 18,000 square centimeters of skin, which accounts for around 18 percent of his body weight.[10] With these statistics it is little wonder that the skin provides us with so much data about our immediate world. We rely on our sense of touch to confirm the inputs of our other senses. We tend to trust our sense of touch in ways that we don't do with our other senses: "I couldn't believe my eyes," "I could hardly believe what I was hearing." A painting titled *The Incredulity of Saint Thomas* by Caravaggio (1571–1610) depicts Apostle Thomas, one of the twelve disciples, touching the body of the resurrected Christ to confirm physically that the Messiah had indeed risen from the dead by feeling the wounds of the crucifixion nails.[11] The incident earned Saint Thomas the nickname Doubting Thomas.

Our shopping experiences are enhanced by our sense of touch. Apart from the physical sensations literally at our fingertips, our entire skin communicates to us.[12] Called haptic information processing (from the Greek word *haptikos*, meaning "the ability to hold or touch"), this innate attribute of our bodies enables us to engage with our immediate environment. Researchers and specialists in this field separate haptic from tactile processing and passive from active use of these sense mechanisms.[13]

A sense of touch enables shoppers to gather and assess information about the product they intend to purchase (or not). Different shoppers use their sense of touch in different ways. Some shoppers are minimalist in their approach, touching products solely to place them in their shopping cart. For others, touch has a wider range of uses, for example, to gauge the size, temperature, ripeness, weight, or robustness of a product. Most children and many adults are inveterately tactile. Shopkeepers who post a sign informing customers "do not handle or touch the fruit and vegetables" deny their potential customers a useful tool for evaluating produce and may be impeding purchases. Retail specialists say that touch provides a strong motivation for a shopper to buy a product.[14] For many consumers, touching a product supplements available information that may be scarce or potentially misleading.[15] Consumers touch products to seek assurance from their personal experience based on prior knowledge (perhaps muscle memory) of what the product should ideally feel like. A need for touch scale attempts to identify, quantify, and explain the preferences and emphases that different people place on their sense of touch when they go shopping.[16] Our sense of touch (haptic information) influences how we feel about a shopping environment and what we choose to buy in that environment.[17] It is likely that allowing consumers to touch products extends the length of time that a consumer spends in a shop.

Another of the shopping innovations to be found at Le Bon Marché was the fixed price of the goods as signified by pricing labels on the merchandise. This was not merely a marketing gimmick to increase profits by preventing customers from negotiating a lower price (haggling), but represented a steep change in consumer shopping behavior. At that time, the predominant form of buying and selling was to haggle.[18]

When the customers can see the prices of the merchandise they gain independence from the shop assistants. The floor space allows customers to move from one fixture to another to compare prices and quality. While not yet self-service (several more decades would be needed before that became accepted shopping behavior), the customer was being given the freedom to compare products without the polite (but sometimes insistent) intrusion of the shop assistant. As was conventional at the time, payment and packaging of the chosen product was still the preserve of the shop assistant. As the customer now had more flexibility of movement and choice, so the shop assistants had a degree of flexibility of action and focus. By not being tied to the one customer they were currently serving, shop assistants could spend more time with particular customers (such as regulars, perceived high-worth individuals, or those requiring extra levels of service). At a time when shop work was becoming more of a profession (especially for females), this allowed for a person to gain experience and expertise in service. Famously, some years before registering herself as a designer in 1919, Gabrielle "Coco" Chanel (1888–1971) had opened and run two shops. She opened the first, a millinery boutique, in Paris in 1910 and was by all accounts very successful, the hats being bought by theater companies and fashionable society. She opened her second boutique in the fashionable seaside resort of Deauville on France's southwest coast. The shop sold Chanel-designed clothes and hats that were modeled for customers and townsfolk by her sister and her aunt.[19] Coco Chanel may have been among the first fashion designers to have worked in a shop, but she probably won't be the last.

An interesting contrast of changes in shop layout and service functions can be found in a recent novel in which a series of unfortunate personal circumstances have prevented the protagonist from shopping to buy razor blades for several decades. Entering a chemist's shop in 2011, the modern shop layout is disconcerting to him. The realignment of service from the shop assistant to the customer clearly comes as a shock:

> The last time I had properly gone shopping was back in 1924 or 1925. In those days one could go to a haberdashery or soap shop. To purchase a razor nowadays, one had to frequent a chemist's; Fräulein Krömeier had told me how to get there. Rossmann was the name. Upon arrival I realized that the appearance of the chemist's had changed out of all recognition. Once upon a time there was a counter, and behind this counter were the goods. Although there was still a counter, now it was situated close to the entrance. Behind it was nothing but a window display. The actual goods were stacked on an endless succession of shelves, for every man to help himself. My initial supposition was that there were dozens of sales assistants, all in informal dress. But it turned out that these were the customers, who wandered about collecting their items and then took these to the counter. It was most disconcerting. Rarely had I felt so impolitely treated. It was as if I had been told on the way in to look for the paltry razor blades myself, as the chemist had better things to do.[20]

However, considering the environment and evaluating the changes from the perspective of the shop owner reveals a number of commercial advantages

to the shop layout. In fact, the chemist shop described by the novel's protagonist bears all of the hallmarks of a classic format of a self-service supermarket. A store reconfigures its floor area and shelving to allow customers to
meander freely through the aisles so that they can handle displayed goods
and compare the range of products in terms of price, packaging, weight, size,
freshness, and other attributes to meet the customer's needs. Ostensibly to aid
the customer, the store provides baskets and wheeled trolleys in which customers can place their goods. The supermarket shopping trolley, invented in
1937 (and patented in 1938) by American supermarket owner Sylvan Nathan
Goldman (1898–1984), helped self-service stores solve their business problem
of how to move goods from the open shelves to the checkout counter[21] when
customers have access to the shelves and the store operates with a reduced
number of serving staff. The shopping trolley is a further example of how
service shifted from the serving staff to the customer. And, unlike a shopping basket, which a customer needs to carry and which increases in weight
with each additional selected item, a trolley doesn't reveal the weight of the
goods being bought. It is thus conceivable that the customer is likely to buy
more. Once a customer decides on the purchases, the checkout system and
process is located near the exit.[22] In the novel our newly arrived protagonist
is not slow to understand the business economics that underpin the modern
design of the chemist's shop:

> Gradually, however, I grasped the logistics inherent here. There were
> indeed a number of advantages to this system. First of all the chemist
> could make large sections of his sales depot accessible, thus affording
> him greater selling space. Furthermore, it was obvious that one hundred
> customers would serve themselves quicker than ten or even twenty
> shop assistants could have done. And, last but not least, one could save
> money by dispensing with these shop assistants. The benefits were
> crystal clear.[23]

Where shop assistants and customers share responsibilities for service
provision, the customer will have predictable reservations and, to a certain
extent, fears.[24] Unlike over-the-counter service transactions, in self-service
environments the customer needs to navigate a free-flowing physical environment in which merchandise, other customers, and shopping equipment
such as trolleys intrude into a shopper's gangway. Physical movement may
be further inhibited by other shoppers' unfamiliarity with the store layout
and their being caught up in the attraction of browsing.

Retailing in the past was dominated by the moment of truth in the meeting
of the staff and the customer. The retail shop had many times direct contact
with producers. But the storage and the packaging were to a large extent
done in the shop in close contact and cooperation with the customer. But
in the 1950s we could see the beginning of a rapid change in this process,

particularly in the United States. About a decade later new methods of retailing started to occur also in Europe and very early in Sweden.

The main issue was a redistribution of tasks between the shop assistant and the buyer. This was the beginning of the first science of automation in the process of retailing. Much earlier we saw the introduction of ten-key keyboards for the cashiers to enter the price and some special codes of the goods to be purchased. At the time, there was quite a heavy discussion in retailing of whether the ten-button keyboard was an improvement or not. Many of the cashiers thought that a full keyboard from companies such as NCR and Hugin would be much more efficient for the task. To enter the sum of, for example, $100 was much faster than to enter 1-0-0 on a ten-button keyboard.

One of the current authors (TI) was involved in evaluating this paradigm shift in retailing. As history now reveals, this paradigm shift became much larger than was apparent at the time. It became much larger than merely the change of keyboard style in retailing. It was the start of a dramatic change in retailing. During the 1960s, the whole concept of retailing started to change, from the design of keyboards to the first attempts to automation in the process of handling the finances, stock keeping, storage facilities, and other components of the retailing process.

In the United States the changes preceded by ten years or more the changes we saw in Europe. However, the co-op chain of retailers was involved very early in this process of change. This retail group took a leading role in Europe. A key concept was the introduction of special checkout systems. Another important component was the introduction of the ten-button keyboards. In a few chains this new system was connected to local computer systems. In the case of the co-op chain these were on a regional basis connected to a kind of mini-computer bought from the Saab aircraft manufacturer. This was probably one of the first examples of a mini-computer in an administrative application. An interesting point from an ergonomics point of view was the involvement of the co-op's ergonomics department and department of environmental studies. Already from the beginning the design of the new process integrated a systems ergonomics approach. As a part of this systems ergonomics approach, the ergonomics department developed a systemized allocation of function according to the model of Singleton in the UK and Ivergård of Sweden. This allocation of functions made it possible to define what tasks should be carried out by people and what tasks should be handed over to automation.

Very early in this process the ergonomists of the co-op understood that the new work structure in the checkout system would create a lot of physical problems. Remedies needed to be introduced very early in the design. From an ergonomics point of view, the workplace in the form of a checkout unit was based on a sitting/working posture where all of the physical work could be handled easily by the cashier. The physical handling of goods (which can be up to many thousands of kilograms per day) needed to be simplified. The workload on the cashiers could be reduced to some extent by the use of two sets of conveyor belts. Evaluation of these checkout systems indicated a high level of productivity improvements. However, the workload was still too heavy and constituted a health risk for the cashier.

The Swedish co-op company Hugin had in parallel, but independently, been developing an automatic reading device for inputting information about lotteries and sports events such as football matches and horseracing. The co-op's ergonomics department was asked to make an evaluation of the Hugin reading system to look into the possibility of introducing other application areas of this optical reading device. The co-op's ergonomist immediately understood the possibility of using this form of reading device as a complement or an alternative to the new keyboards. An internal designer at the co-op's design department had already designed circular codes that could be used for optical registration of information. These optical readers would, in other words, be suitable for incorporating into the checkout system. Using these devices, the checkout assistant could read information from various angles. However, in the United States, a company had already been developing a standard for reading bar codes. The co-op designers tried to persuade their U.S. colleagues of the advantages of a circular bar code reading device. However, the Swedish co-op was too small to be able to influence the selection of circular codes in a bar code system. Over the past forty years, bar code readers have been developed and are rather competitive to circular codes. Most likely in the near future we will see other types of radio-transmitted codes, and bar codes or circular codes will have mostly outlived their usefulness.

At this time, there is occurring a reallocation of checkout systems. Instead of allowing the cashier to identify the purchased items, this function is being transferred to the customer. Nowadays, in many supermarkets, the customer uses a handheld bar code reader. In this way, the workload of the cashier has been reduced to a large extent. Their tasks now function more as a kind of help desk. In the near future we will probably see a greater acceptance of self-service electronic identifications of what has been purchased by the customer. Successively, one is also building up new types of computer systems to support the purchasing process and also to give shopping advice to the consumer. A future development would probably be to send basic commodities directly to the consumer. The consumer will only add on extra items over and above the base commodities. Most likely, the future computerized retailing system will probably also provide additional advice and information to the customer. This is obviously also a way to create a competitive advantage for particular shops. Hopefully, this type of system could also provide new opportunities for supermarkets of the future to provide new services and new opportunities and moments of truth. An interesting issue in relation to this is, of course, if this can be a method for people to have a higher level of health awareness. In this way, future retailers can be a part of preventive health in their society.

Endnotes

1. Interesting analyses and discussions can be found in Kathryn A. Morrison (2003), *English Shops and Shopping: An Architectural History*, New Haven, CT: Yale University Press; William Lancaster (1995), *The Department Store: A Social History*, Leicester, UK: University of Leicester Press; Jan Whittaker (2011), *The Department Store: History, Design, Display*, London: Thames and Hudson Publishers; also see John Benson and Laura Ugolini (eds.) (2003), *A Nation of Shopkeepers: Five Centuries of British Retailing*, London: I.B. Tauris & Co. Ltd.; Adrian R. Bailey, Gareth Shaw, Andrew Alexander, and Dawn Nell (2010), Consumer Behaviour and the Life Course: Shopper Reactions to Self-Service Grocery Shops and Supermarkets in England c. 1947–1975, *Environment and Planning A: International Journal of Urban and Regional Research*, 42(6), 1496–1512.

2. See the description in Erika Diane Rappaport (2000), *Shopping for Pleasure: Women in the Making of London's West End*, Princeton, NJ: Princeton University Press, p. 16.

3. See the discussion in Susan Porter Benson (1988), *Counter Cultures: Saleswomen, Managers, and Customers in American Department Stores 1890–1940*, Champaign: University of Illinois Press, pp. 12ff.

4. For descriptions of the social history and service innovations in retailing see, for example, Michael Barry Miller (1994), *The Bon Marché: Bourgeois Culture and the Department Store, 1869–1920*, Princeton, NJ: Princeton University Press; Gordon Honeycombe (1984), *Selfridges: Seventy-Five Years: The Story of the Store 1909–84*, London: Park Lane Press; Erika D. Rappaport (1999), *Shopping for Pleasure: Women in the Making of London's West End*, Princeton, NJ: Princeton University Press; Jan Whittaker (2011), *The Department Store: History, Design, Display*, London: Thames and Hudson Publishers.

5. See the evidence from empirical research in Joann Peck and Terry L. Childers (2003), To Have and to Hold: The Influence of Haptic Information on Product Judgments, *Journal of Marketing*, 67(2), 35–48; Joann Peck and Jennifer Wiggins (2006), It Just Feels Good: Customers' Affective Response to Touch and Its Influence on Persuasion, *Journal of Marketing*, 70(4), 56–69; Joann Peck and Jennifer Wiggins-Johnson (2011), Autotelic Need for Touch, Haptics and Persuasion: The Role of Involvement, *Psychology of Marketing*, 28(3), 222–239. Also see Jacob Hornik (1991), Shopping Time and Purchasing Behavior as a Result of In-Store Tactile Stimulation, *Perceptual and Motor Skills*, 73, 969–970.

6. Interesting scientific analyses and discussions of the human sense of touch can be found in Ashley Montagu (1986), *Touching: The Human Significance of the Skin*, New York: Harper & Row Publishers; Alberto Gallace and Charles Spence (2010), The Science of Interpersonal Touch: An Overview, *Neuroscience and Biobehavioral Reviews*, 34, 246–259; Alberto Gallace and Charles Spence (2014), *In Touch with the Future: The Sense of Touch from Cognitive Neurosciences to Virtual Reality*, Oxford: Oxford University Press.

7. See relevant discussions in Joann Peck (2010), Does Touch Matter? Insights from Haptic Research in Marketing, in Aradhna Krishna (ed.), *Sensory Marketing: Research on the Sensuality of Products*, New York: Routledge, pp. 17–32.

8. Ashley Montagu (1986), *Touching: The Human Significance of the Skin* (3rd ed.), New York: Harper & Row.

9. Ashley Montagu (1986), *Touching: The Human Significance of the Skin* (3rd ed.), New York: Harper & Row, p. 3.

10. Ashley Montagu (1986), *Touching: The Human Significance of the Skin* (3rd ed.), New York: Harper & Row.

11. The relevant passage in the scriptures can be found in the Authorized (King James) Version of the Bible, John 20:24–29. The Caravaggio painting *The Incredulity of Saint Thomas* can be seen in the Schloss Sanssouci Palace in Potsdam, Brandenburg, Germany.

12. Alberto Gallace, Hong Z. Tan, and Charles Spence (2007), The Body Surface as a Communication System: The State of the Art after 50 Years of Research, *Presence: Teleoperators and Virtual Environments*, 16, 655–676.

13. See explanations in Jack M. Loomis and Susan J. Lederman (1986), Tactile Perception, in Kenneth R. Boff, Lloyd Kaufman, and James Peringer Thomas (eds.), *Handbook of Perception and Human Performance: Cognitive Processes and Performance* (vol. II), New York: John Wiley & Sons, Chapter 31; Roberta L. Klatzky and Susan J. Lederman (2002), Touch, in A.F. Healy and R.W. Proctor (eds.), *Experimental Psychology* (vol. 4), Hoboken, NJ: Wiley, pp. 147–176.

14. There is a wide literature on tactile experiences in shopping. See, for example, Deborah Brown McCabe and Stephen M. Nowlis (2003), The Effect of Examining Actual Products or Product Descriptions on Consumer Preference, *Journal of Consumer Psychology*, 13(4), 431–439; Joann Peck and Jennifer Wiggins (2006), It Just Feels Good: Customers' Affective Response to Touch and Its Influence on Persuasion, *Journal of Marketing*, 70, 56–69; Aradhna Krishna (ed.) (2010), *Sensory Marketing: Research on the Sensuality of Products*, New York: Routledge.

15. See Joann Peck and Jennifer Wiggins (2006), It Just Feels Good: Customers' Affective Response to Touch and Its Influence on Persuasion, *Journal of Marketing*, 70, 56–69. Also see Brenda Soars (2009), Driving Sales through Shoppers' Sense of Sound, Sight, Smell and Touch, *International Journal of Retail and Distribution Management*, 37(3), 286–298.

16. See relevant discussions in Joann Peck and Terry L. Childers (2003), Individual Differences in Haptic Information Processing: The 'Need for Touch' Scale, *Journal of Consumer Research*, 30, 430–442; Bianca Grohmann, Eric R. Spangenberg, and David E. Sprott (2007), The Influence of Tactile Input on the Evaluation of Retail Product Offerings, *Journal of Retailing*, 83(2), 237–245.

17. See discussions and examples in Aradhna Krishna (2010), An Introduction to Sensory Marketing, in Aradhna Krishna (ed.) (2010), *Sensory Marketing: Research on the Sensuality of Products*, New York: Routledge, pp. 1–16. Also see Joann Peck (2010), Does Touch Matter? Insights from Haptic Research in Marketing (pp. 17–32), and Terry L. Childers and Joann Peck (2010), Informational and Affective Influences of Haptics on Product Evaluation: Is What I Say How I Feel? (pp. 63–72) in the same volume.

18. See Gary Davies (1999), The Evolution of Marks and Spencer, *Service Industries Journal*, 19(3), 62.

19. See relevant chapters in Axel Madsen (2009), *Coco Chanel: A Biography*, London: Bloomsbury Publishing; Justine Picardie (2010), *Coco Chanel: The Legend and the Life*, New York: HarperCollins Publishers.

20. Timur Vermes (2014), *Look Who's Back: A Merciless Satire*, London: Maclehose Press (an imprint of Quercus), pp. 334–335, translated from German by Jamie Bulloch. The German title is *Er is wieder da*. The comic device of the novel is that Adolf Hitler (1889–1945) survived WWII and "wakes up" in 2011 in Berlin, where he becomes a celebrity on a television alternative comedy show and on YouTube for his biting critiques of current-day life and politics (yes, this is irony). The novel belongs to the canon of satirical novels from *Gulliver's Travels* by Jonathan Swift (1667–1745), *The Good Soldier Švejk* by Jaroslav Hašek (1883–1923), *Animal Farm* by George Orwell (1903–1950), and *Catch-22* by Joseph Heller (1923–1999). With all of these classic novels, *Look Who's Back* places the protagonist into an environment where the inhabitants behave in ways that are strange and seemingly unfathomable.

21. For a brief history of the shopping trolley, see Walter Y. Oi (2004), The Supermarket: An Institutional Innovation, *Australian Economic Review*, 37(3), 337–342; Catherine Grandclément (2006), Wheeling Food Products around the Store ... and Away: The Invention of the Shopping Cart, 1936–1953, CSI Working Papers Series 006, paper prepared for the Food Chains Conference: Provisioning, Technology, and Science, Wilmington, CT, November 2–4. Also see Terry P. Wilson (1978), *The Cart That Changed the World: The Career of Sylvan N. Goldman*, Norman: University of Oklahoma Press.

22. See discussions in Gareth Shaw, Louise Curth, and Andrew Alexander (2004), Selling Self-Service and the Supermarket: The Americanization of Food Retailing in Britain, 1945–1960, *Business History*, 41(4), 568–582. Also see James M. Mayo (1993), *The American Grocery Store: The Business Evolution of an Architectural Space*, Westport, CT: Greenwood Press.

23. Timur Vermes (2014), *Look Who's Back: A Merciless Satire*, London: Maclehose Press (an imprint of Quercus), p. 335.

24. See the discussions in Andrew Alexander, Dawn Nell, Adrian R. Bailey, and Gareth Shaw (2009), The Co-Creation of a Retail Innovation: Shoppers and the Early Supermarkets in Britain, *Enterprise and Society*, 10(3), 529–558.

12

The Future of Service Excellence through People and Technology

It's tough to make predictions, especially about the future.

Among the witticisms attributed to the former manager of the New York Yankees Lawrence Peter "Yogi" Berra is "it's tough to make predictions, especially about the future."[1] This shrewd advice encourages us to be circumspect in what we write. Rather than make outright prophesies, we consider the future by revisiting the past. One person who has been more prescient than most about the future of technology is John Diebold (1926–2005). A graduate of Harvard Business School, where he earned a distinction, John Diebold was author of such best-selling books as *Automation* (written in 1952 when he was aged twenty-six and one year fresh out of Harvard), *Making the Future Work* (1964), and *Managing Information: The Challenge and the Opportunity* (1985).[2] In an article published in 1965, Diebold described the "threshold of 'information revolution' that will affect the practice of management in ways that our conventional notions of computers can only hint at."[3] To help pay for his studies, Diebold took a low-paying job in a consulting company (which no classmates wanted). Diebold later bought the company. With his first book, John Diebold is acknowledged to have originated many of the technological concepts that are now commonplace, and is acknowledged to have originated the word *automation* as it is used nowadays. John Diebold's consulting company focused on helping organizations understand and appreciate the business and social benefits of information technology (IT).

In this book our motifs have been people, service, and technology. Our recurrent themes have been the concept of the moment of truth and ergonomics/human factors. In this chapter we revisit our themes and offer perspectives for the future. As noted in earlier chapters, the moment of truth is the interface between service provider and customer. Richard Normann estimated that a large company in the service sector experiences tens of thousands of moments of truth every day of its operations.[4] At each moment of truth both service provider and customer have opportunities to assess service quality. Subsequent actions of the customer (buying or walking away) give an indication of some aspects of service quality. We are not entirely convinced that after a service encounter service providers take actions of similar decisiveness. From our personal experiences and from talking with others, we note that a vast majority of service providers do not evaluate the quality of a service encounter in the same manner as their customers. Furthermore, from our own experiences of service and our observations of service offered to others, we see that the vast majority of service providers not only neglect to take this opportunity, but also regard such opportunities as an intrusion into the ways they do business.

The Moment of Truth from an Ergonomics Perspective

In designing people/technology systems, the actual design process is critical in creating a successful moment of truth in the meeting between the technology and people. Ideally, direct participation is needed from the intended users.[5] It is generally accepted that involving users in product design can improve the design of products.[6] Similarly, information from users' experience of service can helpfully contribute to improve service design.[7] Admittedly, on many occasions this might be difficult, as the expected users may not be available at the design stage of the process. This is especially so when designing generic technological systems intended for use by users with diverse attributes. To address this deficiency, a number of work tools have been developed. In a macro-perspective on ergonomics and design of new people/technology service systems, it is necessary to analyze the intended user group and their available skills and competencies. This has to be related to the demand for skills and competencies in achieving certain functions or objectives. This theoretical model for defining an optimal level of automation for supplementing technology will rely on the different criteria mentioned before and related to ergonomic design work.

In other chapters in this book we describe concepts and practices of the moment of truth. Traditionally in these management theories, the setting is a meeting of personnel on one side and the consumers/users on the other. Apart from these two participants, service organizations have a leader, support staff, and support processes (such as quality standards). In this new ergonomic approach to service management the use of technology is included to a greater or lesser extent to give expanded parameters for the moment of truth. In this context, principles of ergonomics become a vital component for analysis, design, and use. Normally the available skills and knowledge among the intended user groups are much more widespread than what is needed for a certain task. But at the same time, the intended user groups (in the current context users can be employees or customers) may not necessarily possess the right skills and competencies needed to create excellence in service. Technology, for example, in the form of different types of information, can supplement these missing skills and competencies.

Service through Automation: A New Era Takes Shape

Automation is not new. For millennia humankind has sought to replace human energy and effort with simple tools and subsequently with machinery. It is said that the invention of the wheel in and of itself was not that world changing. After all, elliptical rocks and pebbles roll down gradients all the time when displaced by footfalls. What is interesting is the next stage of development, when someone conceived that circular rocks could be matched in pairs or in sets of four connected by a rigid pole and a platform added. Then the shapes that had rolled down a gradient of their own accord could now partly replace the burdens of humans or their pack animals. From this relatively early experience of harnessing modest forms of technology, a history of humankind can plot technological development.[8] In the so-called information age, a history of human development with computers is no less fascinating.[9]

Throughout this book we have focused on the interactions between people and technology for the provision of customer services. Service mediated by technology calls for an expanding interpretation of the moment of truth. When technologies aid and support the service encounter (service task, service standards, and service delivery), there is a different framework for service management. A customer's role in a service encounter will need to be reframed where there is no apparent human service provider. While for some components of the service technology may be faster (comparing process, for example), there will be a noticeable lack of human responsiveness to the emotional content of the service encounter. As we know from personal experiences, gaining service solely from technology (such as online inquiries or commercial transactions) can strain the patience of the customer.

Below we describe several possible emerging scenarios of people and technology interactions and describe (in brief) their different roles. We begin from a simple form (traditional human-human service provision) and develop to more advanced levels of technology as an agent-actor in service provision.

The conventional human-to-human provision of service arguably accounts for the overwhelming structure of service encounters throughout the world. Buyers and sellers in traditional market environments taking place daily on all continents engage in this form of service encounter. A possible exception may be Antarctica, where scientific researchers replenish their supplies through radio and forms of information and communication technology (ICT). At the most fundamental configuration the human provider of services (whether frontline face-to-face or from a back office) engages with other humans to offer, transact, and deliver services. The service may be physical (traditional market transactions, healthcare, and personal services) or intellectual (information services such as giving directions to a location, educational services). Increasingly, services may be virtual and delivered to consumers through technological systems and processes. When technologies aid or deliver services to customers, the technology interface (which the customer experiences) needs to be user-friendly. Technology tends to substitute process efficiency for emotional content. At present, technology is emotion-free and efficiency (e.g., speed, consistency, repeated routines) replaces human speed response times and possible error through boredom with repetition. When humans are expected to approach the predictability of machines, employers tend to produce instruction manuals with a range of acceptable responses to be made by the employee in different service encounter eventualities with the customer. Often rote responses to customers by service providers do little to make a positive impression of the service. In the absence of a human service provider, and as current mainstream technologies are emotion-free, technology interfaces with customers need careful design and management. Apart from speed of response and faster transaction times, it would be of little value to replace a human service provider with a machine.

A slightly different situation is evident when people engage in physical activity with technological support. Here the technology may supplement or enhance human physiognomies. Technologies can give added strength or speed (e.g., various items of construction equipment) or reduce the threats of injury (e.g., equipment used in fighting fires, bomb disposal, or mine clearing). The purpose of these technologies is for biofeedback to improve the human operator's current and actual performance. In the hands of highly skilled and professional operators, the human-machine partnership can be impressive, with each augmenting the strengths of the other. In this context the technology provides service to the human customer, although the situation may not be immediately recognizable as a service encounter as such. Nonetheless, the situation shows some features that are useful in service design. Ideally, the service encounter is co-produced with both the service provider and the customer contributing equally (or near equally) to shaping the service encounter. The total service encounter is shaped for mutual benefit (a win-win scenario for each participant).

A greater role of technology in service delivery to the customer is when the service user reacts physically to responses of technology signals, for example, when participating in online activity. In this sense, the human customer acts as a role player in a technological (web-based or Internet) preprogrammed experience. The human participates in social role playing mediated by the ICT. This example gives the user benefits that are mainly virtual and possible experiences of satisfaction. Simulators are used for skills training (such as learning to fly an aircraft) and are sometimes also used for personality and social training.

The Customer as Co-Producer of Value in the Service Encounter

Work by Robert F. Lusch and his colleagues focuses on service-dominant (S-D) logic, one of which foundation premises (FPs) is that the customer is always a co-producer in the production of value for the service encounter.[10] The role of co-producer in a service encounter extends to the internal customer in the service organization.[11] In our earlier examples and discussions we envisage the internal customer to include personnel responsible for service standards and service delivery (see Chapter 1). From the side of the service provision, it is suggested that the whole organization is responsible for delivering service.[12] Jan Carlzon, the astute president of SAS, demonstrated this mind-set when he gave all SAS employees a marketing function. In the case of SAS, this ensured that all employees were committed to delivering quality at the 50 million moments of truth identified by Jan Carlzon in his organization.[13]

When the personal nature of service so engages the customer as to elicit emotional involvement, there is a strong likelihood that such a customer will become a regular user of the service and possibly a loyal advocate of the service to others. Ideally, then, service-providing organizations should aim to provide service excellence to ensure that customers climb the loyalty ladder toward its summit. Using an example of a technology product, service-providing organizations need to design their service in a way similar to that in which Apple designs its products. At every Apple product launch around the world customers are prepared to wait in line to become customers. Indeed, it is more than a product launch and more like a cultural event.[14] While some skeptics may say this is a result of very effective marketing, nonetheless, customers of Apple products, but especially the iPad and iPhone, say they cannot lead their lives without these products.[15] Only rarely is it possible to hear such a customer response about a service.

Figure 12.1 shows that in the delivery of service an organization needs to engage with both internal and external resources in order to ensure the delivery of smooth service to the customer.

Co-creation of service value depends on components that include the service offering (customer needs at the time of the service), the value proposition (the added value gained by the customer in the service), conversation and dialogue (ongoing between the service provider and the customer), and the value network and processes (contributory agents and delivery systems for the service). According to Lusch et al. (2007), "S-D logic superordinates service (the process of providing benefit) to products (units of output that are sometimes used in the process)."[16] This gives the service organization a "service-centered view that is customer oriented and relational," and its resources are employed in serving the customer and building relationships.[17]

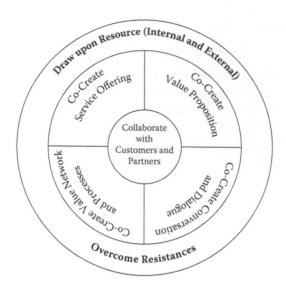

FIGURE 12.1
Service-dominant marketing. (From Robert F. Lusch, Stephen L. Vargo, and Matthew O'Brien (2007), Competing through Service: Insights from Service-Dominant Logic, *Journal of Retailing*, 83(1), 7.)

We began this chapter with the notion that making predictions about the future can be tough. We can state with almost 100 percent certainty that technology will play an increasingly larger part in this future. In 1965, Gordon E. Moore, then CEO of Fairchild Semiconductors and later the co-founder of Intel, stated that the number of transistors it is possible to place on an integrated circuit seemed to be doubling about every two months.[18] Moore's law, as it has since become known, has developed into a general law that states that technology doubles in capacity and halves in price about every twenty-four months.[19] Anyone with a personal electronic device might justifiably say that nowadays the pace of technological innovation is accelerating.

Will the pace of technological change influence the provision of service? No doubt. This trend has been noticeable for some time. Arguably, a greater influence on service quality is people's increasing willingness to accept technology as an important feature of their lives. High levels of technology acceptance will ensure that customers are ahead of service providers in how they wish to compare, assess, order, and use the services they need. When everyone seems to be "connected" most of the time, service has to deliver values and benefits that customers not only want, but also expect. Technology not only connects businesses to customers (actual and potential), but also connects customers to the wider world, and more importantly, the wider world of competitors' customers.

Bill Shankley (1913–1981), manager of Liverpool Football Club from 1959 to 1974, said, "Some people believe football is a matter of life and death.

I'm very disappointed with that attitude. I can assure you it is much, much more important than that." Organizations whose business has service at its core and whose activities focus mainly on service design, management, and delivery are likely to find increasingly that, for survival, service is a matter of life or death.

Endnotes

1. Lawrence Peter "Yogi" Berra (1925–) spent most of his baseball career from 1946 until 1965 with the New York Yankees, where he played in 100 major league games annually. As a player, coach, or manager, Berra appeared in twenty-one World Series. Retired as a player after the 1963 World Series, Berra was then hired to manage the team. In 1972 he was elected to the Baseball Hall of Fame.
2. *Automation* (1952), published by Van Nostrand Publishers, New York (reissued in 1983 by the American Management Association); *Making the Future Work: Unleashing Our Powers of Innovation for the Decades Ahead* (1964), published by Simon & Shuster, New York; *Managing Information: The Challenge and the Opportunity* (1985), published by Amacom Books, New York.
3. John Diebold (1965), What's Ahead in Information Technology, *Harvard Business Review*, 43, 76–82.
4. Richard Normann (2002), *Service Management: Strategy and Leadership in Service Business* (3rd ed.), Chichester, UK: John Wiley & Sons, p. 21.
5. Gavriel Salvendy (ed.) (2013), *Handbook of Human Factors and Ergonomics*, Hoboken, NJ: John Wiley & Sons.
6. Myung Hwan Yun, Sung H. Han, Sang W. Hong, and Jongseo Kim (2003), Incorporating User Satisfaction into the Look-and-Feel of Mobile Phone Design, *Ergonomics*, 46(13–14), 1423–1440.
7. See Stanislav Karapetrovic (1999), ISO 9000, Service Quality and Ergonomics, *Managing Service Quality*, 9(2), 81–89; Teresa A. Swatz and Dawn Iacobucci (eds.) (2000), *Handbook of Services Marketing and Management*, Thousand Oaks, CA: Sage Publications.
8. See, for example, Arnold Pacey (1990), *Technology in World Civilization: A Thousand Year History*, Cambridge, MA: MIT Press.
9. See, for example, Brad A. Myers (1998), A Brief History of Human-Computer Interaction Technology, *Interactions*, March–April, pp. 44–54. Also see Melvin Kranzberg and Carroll W. Pursell Jr. (1967), *Technology in Western Civilization: The Emergence of Modern Industrial Society Earliest Times to 1900* (vol. 1), Oxford: Oxford University Press; Melvin Kranzberg and Carroll W. Pursell Jr. (1967), *Technology in Western Civilization: Technology in the Twentieth Century* (vol. 2), Oxford: Oxford University Press; Melvin Kranzberg (1995), Technology and History: Kranzberg's Laws, *Bulletin of Science, Technology and Society*, 15, 5–13.
10. Stephen L. Vargo and Robert F. Lusch (2006), Evolving a New Dominant Logic for Marketing, in Robert F. Lusch and Stephen L. Vargo (eds.), *The Service-Dominant Logic of Marketing: Dialog, Debate and Directions*, Armonk, NY: M.E. Sharp, p. 18.

11. Evert Gummesson (2008), Extending the Service-Dominant Logic: From Customer Centricity to Balanced Centricity, *Journal of the Academy of Marketing Sciences*, 36, 15–17.
12. See Robert F. Lusch and Stephen L. Vargo (2006), Service-Dominant Logic: Reactions, Reflections and Refinements, *Marketing Theory*, 6, 281–288; Robert F. Lusch, Stephen L. Vargo, and Matthew O'Brien (2007), Competing through Service: Insights from Service-Dominant Logic, *Journal of Retailing*, 83(1), 5–18; Robert F. Lusch and Stephen L. Vargo (eds.) (2006), *The Service-Dominant Logic of Marketing: Dialog, Debate and Directions*, Armonk, NY: M.E. Sharp.
13. Jan Carlzon (1987), *Moments of Truth*, Ballinger Publishing Company, p. 3.
14. Isabel Pedersen (2008), No Apple iPhone? You Must Be Canadian: Mobile Technologies, Participatory Culture and Rhetorical Transformation, *Canadian Journal of Communication*, 33(3), 491–510.
15. Kyle Mickalowski, Mark Mickelson, and Jaciel Keltgen (2008), Apple's iPhone Launch: A Case Study in Effective Marketing, *The Business Review, Cambridge*, 9(2), 283–288.
16. Robert F. Lusch, Stephen L. Vargo, and Matthew O'Brien (2007), Competing through Service: Insights from Service-Dominant Logic, *Journal of Retailing*, 83(1), 6.
17. Robert F. Lusch, Stephen L. Vargo, and Matthew O'Brien (2007), Competing through Service: Insights from Service-Dominant Logic, *Journal of Retailing*, 83(1), 7.
18. Gordon E. Moore (1965), Cramming More Components onto Integrated Circuits, *Electronics*, April 19, pp. 114–117.
19. David C. Brock (ed.) (2006), *Understanding Moore's Law: Four Decades of Innovation*, Philadelphia: Chemical Heritage Foundation.

Index